NATEF Standards Job Sheets

Automatic Transmissions and Transaxles (A2)

Fourth Edition

Jack Erjavec

Ken Pickerill

CENGAGE
Learning·

Australia • Brazil • Canada • Mexico • Singapore • United Kingdom • United States

CENGAGE
Learning®

NATEF Standards Job Sheets
Automatic Transmissions and
Transaxles (A2),
Fourth Edition
Jack Erjavec & Ken Pickerill

VP, General Manager, Skills and Planning:
Dawn Gerrain

Director, Development, Global Product
Management, Skills: Marah Bellegarde

Product Manager: Erin Brennan

Senior Product Development Manager:
Larry Main

Senior Content Developer: Meaghan Tomaso

Product Assistant: Maria Garguilo

Marketing Manager: Linda Kuper

Market Development Manager:
Jonathon Sheehan

Senior Production Director: Wendy Troeger

Production Manager: Mark Bernard

Content Project Management: S4Carlisle

Art Direction: S4Carlisle

Media Developer: Debbie Bordeaux

Cover image(s): © Snaprender/Dreamstime.com

© Shutterstock.com/gameanna

© i3alda/www.fotosearch.com

© IStockPhoto.com/tarras79

© IStockPhoto.com/tarras79

© Laurie Entringer

© IStockPhoto.com/Zakai

© stoyanh/www.fotosearch.com

For product information and technology assistance, contact us at
Cengage Learning Customer & Sales Support, 1-800-354-9706

For permission to use material from this text or product,
submit all requests online at **www.cengage.com/permissions.**
Further permissions questions can be e-mailed to
permissionrequest@cengage.com

Library of Congress Control Number: 2014931170

ISBN-13: 978-1-111-64698-1
ISBN-10: 1-111-64698-8

Cengage Learning
200 First Stamford Place, 4th Floor
Stamford, CT 06902
USA

Cengage Learning is a leading provider of customized learning solutions with office locations around the globe, including Singapore, the United Kingdom, Australia, Mexico, Brazil, and Japan. Locate your local office at:
www.cengage.com/global

To learn more about Cengage Learning, visit **www.cengage.com**
Purchase any of our products at your local college store or at our preferred online store **www.cengagebrain.com**

Notice to the Reader
Publisher does not warrant or guarantee any of the products described herein or perform any independent analysis in connection with any of the product information contained herein. Publisher does not assume, and expressly disclaims, any obligation to obtain and include information other than that provided to it by the manufacturer. The reader is expressly warned to consider and adopt all safety precautions that might be indicated by the activities described herein and to avoid all potential hazards. By following the instructions contained herein, the reader willingly assumes all risks in connection with such instructions. The publisher makes no representations or warranties of any kind, including but not limited to, the warranties of fitness for particular purpose or merchantability, nor are any such representations implied with respect to the material set forth herein, and the publisher takes no responsibility with respect to such material. The publisher shall not be liable for any special, consequential, or exemplary damages resulting, in whole or part, from the readers' use of, or reliance upon, this material.

Printed in the United States of America
Print Number: 02   Print Year: 2021

CONTENTS

NATEF TASK LIST FOR AUTOMATIC TRANSMISSIONS

Required Supplemental Tasks (RST) List

Shop and Personal Safety

1. Identify general shop safety rules and procedures.

2. Utilize safe procedures for handling of tools and equipment.

3. Identify and use proper placement of floor jacks and jack stands.

4. Identify and use proper procedures for safe lift operation.

5. Utilize proper ventilation procedures for working within the lab/shop area.

6. Identify marked safety areas.

7. Identify the location and the types of fire extinguishers and other fire safety equipment; demonstrate knowledge of the procedures for using fire extinguishers and other fire safety equipment.

8. Identify the location and use of eyewash stations.

9. Identify the location of the posted evacuation routes.

10. Comply with the required use of safety glasses, ear protection, gloves, and shoes during lab/shop activities.

11. Identify and wear appropriate clothing for lab/shop activities.

12. Secure hair and jewelry for lab/shop activities.

13. Demonstrate awareness of the safety aspects of supplemental restraint systems (SRS), electronic brake control systems, and hybrid vehicle high-voltage circuits.

14. Demonstrate awareness of the safety aspects of high-voltage circuits (such as high intensity discharge (HID) lamps, ignition systems, injection systems, etc.).

15. Locate and demonstrate knowledge of material safety data sheets (MSDS).

Tools and Equipment

1. Identify tools and their usage in automotive applications.

2. Identify standard and metric designation.

3. Demonstrate safe handling and use of appropriate tools.

4. Demonstrate proper cleaning, storage, and maintenance of tools and equipment.

5. Demonstrate proper use of precision measuring tools (i.e., micrometer, dial-indicator, dial-caliper).

Preparing Vehicle for Service

1. Identify information needed and the service requested on a repair order.

2. Identify purpose and demonstrate proper use of fender covers, mats.

3. Demonstrate use of the three C's (concern, cause, and correction).

4. Review vehicle service history.

5. Complete work order to include customer information, vehicle identifying information, customer concern, related service history, cause, and correction.

Preparing Vehicle for Customer

1. Ensure vehicle is prepared to return to customer per school/company policy (floor mats, steering wheel cover, etc.).

Maintenance and Light Repair (MLR) Task List

A. General
A.1.   Research applicable vehicle and service information fluid type, vehicle service history, service precautions, and technical service bulletins.   Priority Rating 1
A.2.   Check fluid level in a transmission or a transaxle equipped with a dip-stick.   Priority Rating 1
A.3.   Check fluid level in a transmission or a transaxle not equipped with a dip-stick.   Priority Rating 1
A.4.   Check transmission fluid condition; check for leaks.   Priority Rating 2

B. In-Vehicle Transmission/Transaxle
B.1.   Inspect, adjust, and replace external manual valve shift linkage, transmission range sensor/switch, and park/neutral position switch.   Priority Rating 2
B.2.   Inspect for leakage; replace external seals, gaskets, and bushings.   Priority Rating 2
B.3.   Inspect, replace, and align powertrain mounts.   Priority Rating 2
B.4.   Drain and replace fluid and filter(s).   Priority Rating 1

C. Off-Vehicle Transmission and Transaxle Repair
C.1.   Describe the operational characteristics of a continuously variable transmission (CVT).   Priority Rating 3
C.2.   Describe the operational characteristics of a hybrid vehicle drive train.   Priority Rating 3

Automobile Service Technology (AST) Task List

A. General Transmission and Transaxle Diagnosis
A.1.   Identify and interpret transmission/transaxle concern, differentiate between engine performance and transmission/transaxle concerns; determine necessary action.   Priority Rating 1
A.2.   Research applicable vehicle and service information fluid type, vehicle service history, service precautions, and technical service bulletins.   Priority Rating 1
A.3.   Diagnose fluid loss and condition concerns; determine necessary action.   Priority Rating 1
A.4.   Check fluid level in a transmission or a transaxle equipped with a dip-stick.   Priority Rating 1

A.5.   Check fluid level in a transmission or a transaxle not equipped
with a dip-stick.   Priority Rating 1
A.6.   Perform stall test; determine necessary action.   Priority Rating 3
A.7.   Perform lock-up converter system tests; determine necessary action.   Priority Rating 3
A.8.   Diagnose transmission/transaxle gear reduction/multiplication
concerns using driving, driven, and held member (power flow)
principles.   Priority Rating 1
A.9.   Diagnose pressure concerns in a transmission using hydraulic
principles (Pascal's law).   Priority Rating 2

B. In-Vehicle Transmission/Transaxle Maintenance and Repair
B.1.   Inspect, adjust, and replace external manual valve shift linkage,
transmission range sensor/switch, and park/neutral position switch.   Priority Rating 2
B.2.   Inspect for leakage; replace external seals, gaskets, and bushings.   Priority Rating 2
B.3.   Inspect, test, adjust, repair, or replace electrical/electronic
components and circuits including computers, solenoids, sensors,
relays, terminals, connectors, switches, and harnesses.   Priority Rating 1
B.4.   Drain and replace fluid and filter(s).   Priority Rating 1

C. Off-Vehicle Transmission and Transaxle Repair
C.1.   Remove and reinstall transmission/transaxle and torque converter;
inspect engine core plugs, rear crankshaft seal, dowel pins,
dowel pin holes, and mating surfaces.   Priority Rating 1
C.2.   Inspect, leak test, and flush or replace transmission/transaxle
oil cooler, lines, and fittings.   Priority Rating 1
C.3.   Inspect converter flex (drive) plate, converter attaching bolts,
converter pilot, converter pump drive surfaces, converter end play,
and crankshaft pilot bore.   Priority Rating 2
C.4.   Describe the operational characteristics of a continuously
variable transmission (CVT).   Priority Rating 3
C.5.   Describe the operational characteristics of a hybrid vehicle drive train.   Priority Rating 3

Master Automobile Service Technology (MAST) Task List

A. General Transmission and Transaxle Diagnosis
A.1.   Identify and interpret transmission/transaxle concern,
differentiate between engine performance and transmission/transaxle
concerns; determine necessary action.   Priority Rating 1
A.2.   Research applicable vehicle and service information fluid type, vehicle
service history, service precautions, and technical service bulletins.   Priority Rating 1
A.3.   Diagnose fluid loss and condition concerns; determine necessary action.   Priority Rating 1
A.4.   Check fluid level in a transmission or a transaxle equipped
with a dip-stick.   Priority Rating 1
A.5.   Check fluid level in a transmission or a transaxle not equipped
with a dip-stick.   Priority Rating 1
A.6.   Perform pressure tests (including transmissions/transaxles equipped
with electronic pressure control); determine necessary action.   Priority Rating 1
A.7.   Diagnose noise and vibration concerns; determine necessary action.   Priority Rating 2
A.8.   Perform stall test; determine necessary action.   Priority Rating 3
A.9.   Perform lock-up converter system tests; determine necessary action.   Priority Rating 3
A.10. Diagnose transmission/transaxle gear reduction/multiplication
concerns using driving, driven, and held member
(power flow) principles.   Priority Rating 1

A.11.   Diagnose electronic transmission/transaxle control systems using appropriate test equipment and service information.   Priority Rating 1

A.12.   Diagnose pressure concerns in a transmission using hydraulic principles (Pascal's law).   Priority Rating 2

B.   In-Vehicle Transmission/Transaxle Maintenance and Repair

B.1.   Inspect, adjust, and replace external manual valve shift linkage, transmission range sensor/switch, and park/neutral position switch.   Priority Rating 2

B.2.   Inspect for leakage; replace external seals, gaskets, and bushings.   Priority Rating 2

B.3.   Inspect, test, adjust, repair, or replace electrical/electronic components and circuits including computers, solenoids, sensors, relays, terminals, connectors, switches, and harnesses.   Priority Rating 1

B.4.   Drain and replace fluid and filter(s).   Priority Rating 1

C.   Off-Vehicle Transmission and Transaxle Repair

C.1.   Remove and reinstall transmission/transaxle and torque converter; inspect engine core plugs, rear crankshaft seal, dowel pins, dowel pin holes, and mating surfaces.   Priority Rating 1

C.2.   Inspect, leak test, and flush or replace transmission/transaxle oil cooler, lines, and fittings.   Priority Rating 1

C.3.   Inspect converter flex (drive) plate, converter attaching bolts, converter pilot, converter pump drive surfaces, converter end play, and crankshaft pilot bore.   Priority Rating 2

C.4.   Describe the operational characteristics of a continuously variable transmission (CVT).   Priority Rating 3

C.5.   Describe the operational characteristics of a hybrid vehicle drive train.   Priority Rating 3

C.6.   Disassemble, clean, and inspect transmission/transaxle.   Priority Rating 2

C.7.   Inspect, measure, clean, and replace valve body (includes surfaces, bores, springs, valves, sleeves, retainers, brackets, check valves/balls, screens, spacers, and gaskets).   Priority Rating 2

C.8.   Inspect servo and accumulator bores, pistons, seals, pins, springs, and retainers; determine necessary action.   Priority Rating 2

C.9.   Assemble transmission/transaxle.   Priority Rating 2

C.10.   Inspect, measure, and reseal oil pump assembly and components.   Priority Rating 2

C.11.   Measure transmission/transaxle end play or preload; determine necessary action.   Priority Rating 1

C.12.   Inspect, measure, and replace thrust washers and bearings.   Priority Rating 2

C.13.   Inspect oil delivery circuits, including seal rings, ring grooves, and sealing surface areas, feed pipes, orifices, and check valves/balls.   Priority Rating 2

C.14.   Inspect bushings; determine necessary action.   Priority Rating 2

C.15.   Inspect and measure planetary gear assembly components; determine necessary action.   Priority Rating 2

C.16.   Inspect case bores, passages, bushings, vents, and mating surfaces; determine necessary action.   Priority Rating 2

C.17.   Diagnose and inspect transaxle drive, link chains, sprockets, gears, bearings, and bushings; perform necessary action.   Priority Rating 2

C.18.   Inspect, measure, repair, adjust, or replace transaxle final drive components.   Priority Rating 2

C.19.   Inspect clutch drum, piston, check-balls, springs, retainers, seals, and friction and pressure plates, bands and drums; determine necessary action.   Priority Rating 2

C.20.   Measure clutch pack clearance; determine necessary action.   Priority Rating 1

C.21.   Air test operation of clutch and servo assemblies.   Priority Rating 1

C.22.   Inspect roller and sprag clutch, races, rollers, sprags, springs, cages, retainers; determine necessary action.   Priority Rating 2

DEFINITION OF TERMS USED IN THE TASK LIST

To clarify the intent of these tasks, NATEF has defined some of the terms used in the task list. To get a good understanding of what the task includes, refer to this glossary while reading the task list.

adjust	To bring components to specified operational settings.
air test	To use air pressure to determine the proper action of components.
align	To bring components into precise alignment or relative position.
assemble (reassemble)	To fit together the components of a device.
check	To verify a condition by performing an operational or comparative examination.
clean	To rid components of foreign matter for the purpose of reconditioning, repairing, measuring, and reassembling.
confirm	To acknowledge something has happened with firm assurance.
Controller Area Network (CAN)	CAN is a network protocol (SAE J2284/ISO 15765-4) used to interconnect a network of electronic control modules. Some manufacturers began implementing CAN with model year 2003. By model year 2008, the California Air Resources Board (CARB) required the use of CAN on all vehicles.
demonstrate	To show or exhibit the knowledge of a theory or procedure.
determine	To establish the procedure to be used to perform the necessary repair.
determine necessary action	Indicates that the diagnostic routine(s) is the primary emphasis of a task. The student is required to perform the diagnostic steps and communicate the diagnostic outcomes and corrective actions required to address the concern or problem. The training program determines the communication method (worksheet, test, verbal communication, or other means deemed appropriate) and whether the corrective procedures for these tasks are actually performed.
diagnose	To identify the cause of a problem.
differentiate	To perceive the difference in or between one thing to other things.
disassemble	To separate a component's parts as a preparation for cleaning, inspection, or service.
drain	To use gravity to empty a container.
fill (refill)	To bring a fluid level to a specified point or volume.
find	To locate a particular problem, such as shorts, grounds, or opens, in an electrical circuit.
flush	To internally clean a component or system.
identify	To establish the identity of a vehicle or component before service; to determine the nature or degree of a problem.
inspect	(see *check*)
install (reinstall)	To place a component in its proper position in a system.
leak test	To locate the source of leaks in a component or system.
listen	To use audible clues in the diagnostic process; to hear the customer's description of a problem.
locate	Determine or establish a specific spot or area.
maintain	To keep something at a specified level, position, rate, etc.
measure	To determine existing dimensions/values for comparison to specifications.
network	A system of interconnected electrical modules or devices.
on-board diagnostics (OBD)	Diagnostic protocol that monitors computer inputs and outputs for failures.
perform	To accomplish a procedure in accordance with established methods and standards.

perform necessary action	Indicates that the student is to perform the diagnostic routine(s) and perform the corrective action item. Where various scenarios (conditions or situations) are presented in a single task, at least one of the scenarios must be accomplished.
pressure test	To use air or fluid pressure to determine the integrity, condition, or operation of a component or system.
priority ratings	Indicates the minimum percentage of tasks, by area, that a program must include in its curriculum in order to be certified in that area.
remove	To disconnect and separate a component from a system.
repair	To restore a malfunctioning component or system to operating condition.
replace	To exchange a component; to reinstall a component.
research	To make a thorough investigation into a situation or matter.
reset	(see *set*)
select	To choose the correct part or setting during assembly or adjustment.
service	To perform a procedure, as specified in the owner's manual or service manual.
set	To adjust a variable component to a given, usually initial, specification.
test	To verify a condition through the use of meters, gauges, or instruments.
torque	To tighten a fastener to a specified degree or tightness (in a given order or pattern if multiple fasteners are involved on a single component).
verify	To confirm that a problem exists after hearing the customer's complaint or concern, or to confirm the effectiveness of a repair.

CROSS-REFERENCE GUIDES

MLR Task	Job Sheet
B.2	23
B.3	26
B.4	25
C.1	30
C.2	31

AST Task	Job Sheet
A.1	10
A.2	11
A.3	12
A.4	13
A.5	14
A.6	17
A.7	18
A.8	19
A.9	21
B.1	22
B.2	23
B.3	24
B.4	25
B.5	26
C.1	27
C.2	28
C.3	29
C.4	30
C.5	31

MAST Task	Job Sheet
A.1	10
A.2	11
A.3	12
A.4	13
A.5	14
A.6	15
A.7	16
A.8	17
A.9	18
A.10	19
A.11	20
A.12	21
B.1	22
B.2	23
B.3	24
B.4	25
B.5	26
C.1	27
C.2	28
C.3	29
C.4	30
C.5	31

PREFACE

The automotive service industry continues to change with the technological changes made by automobile, tool, and equipment manufacturers. Today's automotive technician must have a thorough knowledge of automotive systems and components, good computer skills, exceptional communication skills, good reasoning, the ability to read and follow instructions, and above average mechanical aptitude and manual dexterity.

This new edition, like the last, was designed to give students a chance to develop the same skills and gain the same knowledge that today's successful technician has. This edition also reflects the changes in the guidelines established by the National Automotive Technicians Education Foundation (NATEF), in 2013.

The purpose of NATEF is to evaluate technician training programs against standards developed by the automotive industry and recommend qualifying programs for certification (accreditation) by ASE (National Institute for Automotive Service Excellence). Programs can earn ASE certification upon the recommendation of NATEF. NATEF's national standards reflect the skills that students must master. ASE certification through NATEF evaluation ensures that certified training programs meet or exceed industry-recognized, uniform standards of excellence.

At the expense of much time and thought, NATEF has assembled a list of basic tasks for each of their certification areas. These tasks identify the basic skills and knowledge levels that competent technicians have. The tasks also identify what is required for a student to start a successful career as a technician.

In June 2013, after many discussions with the industry, NATEF established a new model for automobile program standards. This new model is reflected in this edition and covers the new standards, which are based on three (3) levels: Maintenance & Light Repair (MLR), Automobile Service Technician (AST), and Master Automobile Service Technician (MAST). Each successive level includes all the tasks of the previous level in addition to new tasks. In other words, the AST task list includes all of the MLR tasks plus additional tasks. The MAST task list includes all of the AST tasks, plus additional tasks designated specifically for MAST.

Most of the content in this book are job sheets. These job sheets relate to the tasks specified by NATEF, according to the appropriate certification level. The main considerations during the creation of these job sheets were student learning and program certification by NATEF. Students are guided through standard industry-accepted procedures. While they are progressing, they are asked to report their findings as well as offer their thoughts on the steps they have just completed. The questions asked of the students are thought provoking and require students to apply what they know to what they observe.

The job sheets were also designed to be generic. That is, whenever possible, the tasks can be performed on any vehicle from any manufacturer. Also, completion of the sheets does not require the use of specific brands of tools and equipment. Rather, students use what is available. In addition, the job sheets can be used as a supplement to any good textbook.

Also included are description and basic use of the tools and equipment listed in NATEF's standards. The standards recognize that not all programs have the same needs, nor do all programs teach all of the NATEF tasks. Therefore, the basic philosophy for the tools and equipment requirement is that the training should be as thorough as possible with the tools and equipment necessary for those tasks.

Theory instruction and hands-on experience of the basic tasks provide initial training for employment in automotive service or further training in any or all of the specialty areas. Competency in the tasks indicates to employers that you are skilled in that area. You need to know the appropriate theory, safety, and support information for each required task. This should include identification and use of the required tools and testing and measurement equipment required for the tasks, the use of current reference and training materials, the proper way to write work orders and warranty reports, and the storage, handling, and use of hazardous materials as required by the "Right to Know" law, and federal, state, and local governments.

Words to the Instructor: We suggest you grade these job sheets based on completion and reasoning. Make sure the students answer all the questions. Then, look at their reasoning to see if the task was actually completed and to get a feel for their understanding of the topic. It will be easy for students to copy others' measurements and findings, but each student should have their own base of understanding and that will be reflected in their explanations.

Words to the Student: While completing the job sheets, you have a chance to develop the skills you need to be successful. When asked for your thoughts or opinions, think about what you observed. Think about what could have caused those results or conditions. You are not being asked to give accurate explanations for everything you do or observe. You are only asked to think. Thinking leads to understanding. Good technicians are good because they have a basic understanding of what they are doing and why they are doing it.

Jack Erjavec and Ken Pickerill

AUTOMATIC TRANSMISSIONS AND TRANSAXLES

To prepare you to learn what you should learn from completing the job sheets, some basics must be covered. This discussion begins with an overview of automatic transmissions. Emphasis is placed on what they do and how they work, including the major components and designs of automatic transmissions and their role in the efficient operation of automatic transmissions of all designs.

Preparing to work on an automobile would not be complete without addressing certain safety issues. This discussion covers what you should and should not do while working on automatic transmissions, including the proper ways to deal with hazardous and toxic materials.

NATEF's task list for automatic transmissions and transaxles certification is given, along with definitions of some of the terms used to describe the tasks. This list gives you a good look at what the experts say you need to know before you can be considered competent to work on automatic transmissions.

After the task list is descriptions of the various tools and types of equipment you need to be familiar with. These are the tools you will use to complete the job sheets. They are also the tools NATEF has identified as being necessary for servicing automatic transmissions.

After the tool discussion is a cross-reference guide that shows which NATEF tasks are related to specific job sheets. In most cases there is a single job sheet for each task. Some tasks are part of a procedure, in which case one job sheet may cover two or more tasks. The remainder of the book contains the job sheets.

BASIC AUTOMATIC TRANSMISSION THEORY

Many rear-wheel-drive (RWD) and four-wheel-drive (4WD) vehicles are equipped with automatic transmissions. Automatic transaxles, which combine an automatic transmission and final drive assembly into a single unit, are used on front-wheel-drive (FWD) and some all-wheel-drive (AWD) and RWD vehicles.

An automatic transmission or transaxle selects gear ratios according to engine speed, powertrain load, vehicle speed, and other operating factors. In recent years, the push to increase fuel economy ratings has prompted vehicle manufacturers to increase the number of forward gears available on new vehicles to as many as nine. Multiple gear ratios help match the transmission ratio to the engine. All current transmissions have computer-controlled torque converter clutches and shifting. Based on input data supplied by electronic sensors and switches, the computer sets the torque converter's operating mode, controls the transmission's shifting sequence, and have the ability to regulate transmission fluid pressure.

Torque Converter

Automatic transmissions use a torque converter (Figure 1) to transfer engine torque from the engine to the transmission. The torque converter operates through hydraulic force provided by fluid. The torque converter automatically engages and disengages power from the engine to the transmission in relation to engine rpm. With the engine running at the correct idle speed,

Copyright © 2015 Cengage Learning®.

Figure 1 A torque converter and automatic transmission.

there is not enough fluid flow for power transfer through the torque converter. As engine speed increases, the added fluid flow creates sufficient force to transmit engine power through the torque-converter assembly to the transmission.

The torque converter, located between the engine and transmission, is a sealed, doughnut-shaped unit that is always filled with automatic transmission fluid. A special flex plate is used to mount the torque converter to the crankshaft. The purpose of the flex plate is to transfer crankshaft rotation to the shell of the torque converter assembly. The flex plate also carries the starter motor ring gear. A flywheel is not required because the mass of the torque converter and flex disc acts like a flywheel to smooth out the engine's intermittent power strokes.

A standard torque converter consists of three elements: the impeller, the stator assembly, and the turbine. The impeller assembly is the input member; it receives power from the engine. The turbine is the output member; it is splined to the transmission's input shaft. The stator assembly is the reaction member or torque multiplier. The stator is supported on a one-way clutch, which operates as an overrunning clutch and permits the stator to rotate freely in one direction and lock up in the opposite direction.

The impeller forms one internal section of the torque converter shell. The impeller has numerous curved blades that rotate as a unit with the shell. It turns at engine speed, acting like a pump to start the transmission oil circulating within the torque converter shell.

The impeller is positioned with its back facing the transmission housing, whereas the turbine is positioned with its back to the engine. The curved blades of the turbine face the impeller assembly. The turbine blades have a greater curve than the impeller blades, which helps eliminate oil turbulence between the turbine and impeller blades that would slow impeller speed and reduce the converter's efficiency.

The stator is located between the impeller and turbine. It redirects the oil flow from the turbine back into the impeller in the direction of impeller rotation with minimal loss of speed and force. The side of the stator blade with the inward curve is the concave side. The side with an outward curve is the convex side.

At the rear of the torque converter shell is a hollow hub with notches or flats at one end, set 180 degrees apart. This hub is called the pump drive hub. The notches or flats drive the transmission pump assembly. At the front of the transmission, within the pump housing, is a pump bushing that supports the pump drive hub and provides rear support for the torque converter assembly. Some other transaxles have a separate shaft to drive the pump.

As the impeller rotates, centrifugal force throws the oil outward and upward due to the curved shape of the impeller housing. The faster the impeller rotates, the greater the centrifugal force becomes. Oil thrown outward and upward from the impeller strikes the curved vanes of the turbine, causing the turbine to rotate. (There is no direct mechanical link between the impeller and the turbine.) Oil leaving the turbine is directed out of the torque converter to an external oil cooler and then to the transmission's oil sump or pan.

With the transmission in gear and the engine at idle, the vehicle can be held stationary by applying the brakes. At idle, engine speed is slow. Since the impeller is driven by engine speed, it turns slowly, creating little centrifugal force within the torque converter. Therefore, little or no power is transferred to the transmission.

When the throttle is opened, engine speed, impeller speed, and the amount of centrifugal force generated in the torque converter all increase dramatically. Oil is then directed against the turbine blades, which transfer power to the turbine shaft and transmission.

As the speed of the turbine approaches the speed of the impeller, the coupling point is reached and the turbine and the impeller run at essentially the same speed. They cannot run at exactly the same speed due to slippage between them. The only way they can turn at exactly the same speed is by using a torque converter clutch to mechanically tie them together.

The stator redirects the oil leaving the turbine back to the impeller, which helps the impeller rotate more efficiently. Torque converter multiplication can only occur when the impeller is rotating faster than the turbine.

A lockup torque converter eliminates the 10 percent slip that takes place between the impeller and turbine at the coupling stage of operation. There are many types of lockup torque converters. The lockup piston clutch is the type installed in most automatic transmissions. There are also fully mechanical lockup converters, centrifugal lockup converters, and converters that use a viscous coupling.

Planetary Gears

Nearly all automatic transmissions rely on planetary gear sets to transfer power and multiply engine torque to the drive axles. Compound gear sets combine two simple planetary gear sets so load can be spread over a greater number of teeth for strength and also to obtain the largest number of gear ratios possible in a compact area.

A simple planetary gear set consists of three parts: a sun gear, a carrier with planetary pinions mounted to it, and an internally toothed ring gear or annulus. The sun gear is located in the center of the assembly and meshes with the teeth of the planetary pinion gears. Planetary pinion gears are small gears fitted into a framework called the planetary carrier. The planetary carrier includes a shaft for each of the planetary pinion gears to rotate on. The planetary pinions surround the sun gear's center axis, and the ring gear surrounds them. The ring gear acts like a band to hold the entire gear set together and provide great strength to the unit.

A limited number of gear ratios are available from a single planetary gear set. To increase the number of available gear ratios, gear sets can be added. For many years there were two common designs of compound gear sets: the Simpson gear set, in which two planetary gear sets share a common sun gear, and the Ravigneaux gear set, which has two sun gears, two sets of planetary gears, and a common ring gear. Some automatic

transmissions use two simple planetary units in series. In this type of arrangement, gear set members are not shared; instead, the holding devices are used to lock different members of the planetary units together.

Some late-model six-, seven-, and eight-speed transmissions use the Lepelletier system. The Lepelletier system connects a simple planetary gear set to a Ravigneaux gear set. This design has been around for many years but was difficult to control. Today's electronic technologies have now made it practical. With this design, transmissions can be made with additional forward speeds without an increase in size and weight. In fact, most six speed transmissions are more compact and lighter than nearly all four- or five-speed transmissions.

In this arrangement, the ring gear of the simple gear set serves as the input to the gear sets and, at the same time, the input can be connected to the carrier of the Ravigneaux gear. As engine torque passes through different input gears, it drives a variety of different combinations of gears in the simple and the Ravigneaux gearset. These combinations result in the various forward speed ratios. The ring gear of the Ravigneaux gear set is the output member for the transmission.

In some seven-speed models, the input shaft is always connected to the ring of the simple planetary gear and, in addition, can be connected to the carrier and large sun gear of the Ravigneaux gear. This allows for additional gear combinations. The ring gear of the Ravigneaux gear set still serves as the output.

A few automatic transaxles are based on the design of a manual transmission, that is, they use constant-mesh helical and square-cut gears.

Another unconventional transmission design, the continuously variable transmission (CVT), has no fixed forward speeds. The gear ratio varies with engine speed and temperature. Rather than gears, these transmissions use belts and pulleys. These transmissions also do not have a torque converter; instead, they use a manual transmission–type flywheel with a start clutch. One pulley is the driven member and the other is the drive. Each pulley has a movable face and a fixed face. When the movable face moves, the effective diameter of the pulley changes. The change in effective diameter changes the effective pulley (gear) ratio. A steel belt links the driven and drive pulleys.

Planetary Gear Controls

Certain parts of the planetary gear train must be held while others must be driven to provide the needed torque multiplication and direction for vehicle operation. Bands, brakes, and clutches are the typical planetary gear controls used today.

A band is a braking assembly positioned around a stationary or rotating drum. The band brings a drum to a stop by wrapping itself around the drum and holding it. A servo assembly hydraulically applies the band. Connected to the drum is a member of the planetary gear train. The purpose of a band is to hold a member of the planetary gear set by holding the drum and connecting planetary gear member stationary. The servo assembly converts hydraulic pressure into a mechanical force that applies a band to hold a drum stationary.

In an automatic transmission operation, both sprag and roller overrunning clutches are used to hold members of the planetary gear set. These clutches operate mechanically. An overrunning clutch allows rotation in only one direction and operates at all times. One-way overrunning clutches can be either roller-type or sprag-type clutches.

A multiple–friction disc assembly uses a series of friction discs to transmit torque or apply braking force. The friction discs have internal teeth that are sized and shaped to mesh with splines on the clutch assembly hub. In turn, this hub is connected to a member of the planetary gear set that receives the desired braking or transfer force when the clutch is applied or released.

Multiple-disc clutches are enclosed in a large drum-shaped housing that holds the other clutch components: cylinder, hub, piston, piston return springs, seals, pressure plate, clutch plates, friction plates, and snap rings.

The friction discs are sandwiched between the clutch plates and pressure plate. Friction discs are steel core plates with friction material bonded to either side. The pressure plate has tabs around the outside diameter to mate with the channels in the clutch drum. It is held in place by a large snap ring. Upon engagement, the clutch piston forces the clutch pack against the fixed pressure plate.

Bearings, Bushings, and Thrust Washers

When a component slides over or rotates around another part, the surfaces that contact each other are called bearing surfaces. A gear rotating on a fixed shaft can have more than one bearing surface if it is supported and held in place by the shaft in a radial direction. In addition, the gear tends to move along the shaft in an axial direction as it rotates and is therefore held in place by some other components. The surfaces between the sides of the gear and the other parts are bearing surfaces.

A bearing is placed between two bearing surfaces to reduce friction and wear. In automatic transmissions, sliding bearings are used where one or more of the following conditions prevail: low rotating speeds, very large bearing surfaces compared to the surfaces present, and low use. Rolling bearings are used in high-speed applications, high load with relatively small bearing surfaces, and high use.

Transmissions use sliding bearings made of relatively soft bronze alloy. Many are made from steel with the bearing surface bonded or fused to the steel. Bearings that take radial loads are called bushings and those that take axial loads are called thrust washers.

Since bushings are made of a soft metal, they act like a bearing and support many of the transmission's rotating parts. They are also used to precisely guide the movement of various valves in the transmission's valve body. Bushings can also be used to control fluid flow; some bushings restrict the flow from one part to another, and others are made to direct fluid flow to a particular point or part in the transmission.

Thrust washers are made in various thicknesses. They may have one or more tangs or slots on the inside or outside circumference that mate with the shaft bore to keep them from turning. Some thrust washers are made of nylon or Teflon, which are used when the load is low. Others are fitted with rollers to reduce friction and wear.

Thrust washers normally control free axial movement or endplay. Since some endplay is necessary in all transmissions because of heat expansion, proper endplay is often accomplished through selective thrust washers. These thrust washers are inserted between various parts of the transmission. Thrust washers work by filling the gap between two objects and become the primary wear item because they are made of softer materials than the parts they protect.

Torrington bearings are thrust washers fitted with roller bearings. These thrust bearings are primarily used to limit endplay but also to reduce the friction between two rotating parts. Most often Torrington bearings are used in combination with flat thrust washers to control endplay of a shaft or the gap between a gear and its drum.

Snap Rings

Many different sizes and types of snap rings are used in today's transmissions. External and internal snap rings are used as retaining devices throughout the transmission. Internal snap rings are used to hold servo assemblies and clutch assemblies together. In fact, snap rings are also available in several thicknesses and may be used to adjust the clearance in multiple-disc clutches. Some clutch packs use waved snap rings to provide for smoother clutch applications. External snap rings are used to hold gear and clutch assemblies to their shafts.

Gaskets and Seals

The gaskets and seals of an automatic transmission help to keep the fluid in the transmission and prevent it from leaking out of the various hydraulic circuits. Different types of seals are used in automatic transmissions; they can be made of rubber, metal, or Teflon. Transmission gaskets are made of rubber, cork, paper, synthetic materials, or plastic.

Gaskets are used to seal two parts together or to provide a passage for fluid flow from one part of the transmission to another. Soft gaskets are used when the sealing surfaces are irregular or in places where the surface may distort when the component is tightened into place. A typical location for a soft gasket is the oil pan gasket that seals the oil pan to the transmission case. Oil pan gaskets are typically a composition-type gasket made with rubber and cork. However, some late-model transmissions use a room temperature vulcanizing (RTV) sealant instead of a gasket to seal the oil pan.

As valves and transmission shafts move within the transmission, it is essential that the fluid and pressure be contained within its bore. Any leakage would decrease the pressure and result in poor transmission operation. Seals are used to prevent leakage around valves, shafts, and other moving parts.

O-rings are round seals with a circular cross section. Normally an O-ring is installed in a groove cut into the inside diameter of one of the parts to be sealed. When the other part is inserted into the bore and through the O-ring, the O-ring is compressed between the inner part and the groove. This pressure distorts the O-ring and forms a tight seal between the two parts.

Lip seals are used to seal parts that have axial or rotational movement. They are round so that they fit around shafts, but the entire seal does not serve as a seal; instead, the sealing part is a flexible lip. The flexible lip is normally made of synthetic rubber and shaped so that it is flexed when installed in order to apply pressure at the sharp edge of the lip. Lip seals are used around input and output shafts to keep fluid in the housing and dirt out. Some seals are double-lipped.

A square-cut seal can withstand more axial movement than an O-ring can. Square-cut seals are also round seals but have a rectangular or square cross section. They are designed this way to prevent the seal from rolling in its groove when there is a large amount of axial movement. Added sealing comes from the distortion of the seal during axial movement. As the shaft inside the seal moves, the outer edge of the seal moves more than the inner edge, thereby causing the diameter of the sealing edge to increase, which creates a tighter seal.

Some parts of the transmission do not require a positive seal; some leakage is acceptable. These components are sealed with ring seals that fit into a groove on a shaft. The outside diameter of the ring seals slide against the walls of the bore that the shaft is inserted into. Most ring seals in a transmission are placed near pressurized fluid outlets on rotating shafts to help retain pressure. Three types of metal seals are used in automatic transmissions: butt-end seals, open-end seals, and hook-end seals.

Some transmissions use Teflon seals instead of metal seals. Teflon provides for a softer sealing surface, which results in less wear on the surface that it rides on and therefore a longer-lasting seal. Teflon seals are similar in appearance to metal seals except for the hook-end type. The ends of locking-end Teflon seals are cut at an angle; locking-end seals are often called scarf-cut seals.

Many late-model transmissions are equipped with solid one-piece Teflon seals. Although the one-piece seal requires some special tools for installation, it makes a near-positive seal. These Teflon rings seal much better than other metal sealing rings.

Final Drive Assemblies

The last set of gears in the drive train is the final drive. In most RWD cars, the final drive is located in the rear axle housing. On most FWD cars, the final drive is located within the transaxle. Some FWD cars with longitudinally mounted engines locate the differential and final drive in a separate case that bolts to the transmission.

A transaxle's final drive gears provide a way to transmit the transmission's output to the differential section of the transaxle. Four common configurations are used as the final drives on FWD vehicles: helical gear, planetary gear, hypoid gear, and chain drive. The helical, planetary, and chain final drive arrangements are found with transversely mounted engines. Hypoid final drive gear assemblies are normally found in vehicles with a longitudinally placed engine. The hypoid assembly is basically the same unit as that used on RWD vehicles and is mounted directly to the transmission.

Some transaxles route output torque through two helical-cut gears to a transfer shaft. A helical-cut pinion gear attached to the opposite end of the transfer shaft drives the differential ring gear and carrier.

Other transaxles use a simple planetary gear set for the final drive. In operation, the transmission's output drives the sun gear that, in turn, drives the planetary pinion gears. The pinion gears walk around the inside of the stationary ring gear. The rotating planetary pinion gears drive the planetary carrier and differential case. This combination provides maximum torque multiplication from a simple planetary gear set.

Chain-drive final drive assemblies use a multiple-link chain to connect a drive sprocket, which is connected to the transmission's output shaft, to a driven sprocket, which is connected to the differential case. This design allows for remote positioning of the differential within the transaxle housing. Final drive gear ratios are determined by the size of the driven sprocket compared to the drive sprocket.

Hydraulic System

An automatic transmission uses fluid pressure to control the action of the planetary gear sets. This fluid pressure is regulated and directed to change gears automatically through the use of various pressure regulators and control valves.

The automatic transmission reservoir is the transmission oil pan. Transmission fluid is drawn from the pan and returned to it. The pressure source in the system is the oil pump. The valve body contains control valves to regulate or restrict the pressure and flow of fluid within the transmission. Output devices are the servos or clutches operated by hydraulic pressure.

The transmission pump is driven by the torque converter shell at engine speed. The purpose of the pump is to create fluid flow in the system. Pump pressure is a variable pressure, depending on engine speed from idle to full throttle. The pressure in a transmission's hydraulic system is the result of constant fluid flow from the pump and the valves, pistons, seals, and bushings in the transmission.

Basic Hydraulic Theory

An automatic transmission is a complex hydraulic circuit. Hydraulics is the study of liquids in motion. A fluid is something that does not have a definite shape, therefore liquids and gases are fluids. A characteristic of all fluids is that they will conform to the shape of their container. A liquid also will typically not compress, regardless of the pressure on it. Therefore, liquids are considered non-compressible fluids.

Liquids will, however, predictably respond to pressures exerted on them. Their reaction to pressure is the basis of all hydraulic applications. This fact allows hydraulics to do work.

Over 300 years ago a French scientist, Blaise Pascal, determined that if you had a liquid-filled container with only one opening and applied force to the liquid through that opening, the force would be evenly distributed throughout the liquid. This explains how pressurized liquid is used to operate and control an automatic transmission. A pump pressurizes the transmission fluid and the fluid is delivered to the various apply devices. When the pressure increases enough to activate the apply device, it holds a gear set member until the pressure decreases. The valve body is responsible for the distribution of the pressurized fluid to the appropriate reaction member according to the operating conditions of the vehicle.

Pascal also found he could increase the force available to do work, simply by moving fluid under pressure from a cylinder to a larger one. He determined that force applied to liquid creates pressure or the transmission of force through the liquid. His experiments revealed two important aspects of a liquid when it is confined and put under pressure: the pressure applied to it is transmitted equally in all directions and this pressure acts with equal force at every point in the container.

When a pressure is applied to a confined liquid, the pressure of the liquid is the same everywhere within the hydraulic system. If the hydraulic pump provides 100 psi, there will be 100 pounds of pressure on every square inch of the system. If the system has a piston with an area of 30 square inches, each square inch receives 100 pounds of pressure. This means there will be 3,000 pounds of force applied to that piston. According to Pascal's law, force is equal to the pressure multiplied by the area of piston ($F = P \times A$). The use of the larger piston gives the system a mechanical advantage or increase in the force available to do work.

By changing the size of the pistons in a hydraulic system, force is multiplied, and as a result, low amounts of force are needed to move heavy objects.

Transmission Fluid

The automatic transmission fluid (ATF) circulating through the transmission and torque converter and over the parts of the transmission cools the transmission. The heated fluid moves to the transmission fluid cooler where the heat is removed. As the fluid lubricates and cools the transmission, it also cleans the parts. The dirt is carried by the fluid to a filter where the dirt is removed.

Another critical job of ATF is its role in shifting gears. ATF moves under pressure throughout the transmission and causes various valves to move. The pressure of the ATF changes with changes in engine speed and load.

ATF is also used to operate the various apply devices (clutches and brakes) in the transmission. At the appropriate time, a switching valve opens and sends pressurized fluid to the apply device which engages or disengages a gear.

Valve Body

The purpose of the valve body is to respond to the speed of the engine and the load on it and the drive train, as well as the driver's intentions. The valve body is an aluminum or iron casting with many precisely machined holes and passages that accommodate fluid flow and various valves. Internally, the valve body has many fluid passages.

The purpose of a valve is to start, stop, or direct and regulate fluid flow. In most valve bodies, three types of valves are used: check ball, poppet, and, most commonly, the spool.

The *check ball valve* is a ball that operates on a seat located on the valve body. The check ball operates by having a fluid pressure or manually operated linkage force it against the ball seat to block fluid flow. Pressure on the opposite side unseats the check ball. Check balls and poppet valves can be designed to be normally open, which allows free flow of fluid pressure, or normally closed, which blocks fluid pressure flow. Other applications of the check ball have two seats to check and direct fluid flow from two directions, being seated and unseated by pressures from either source.

Check ball valves can also be used as pressure relief valves to relieve excessive fluid pressure. The check ball is held against its seat by spring tension that is stronger than the fluid pressure. When the fluid pressure overcomes the spring tension, the check ball is forced off its seat, which relieves excess pressure. As soon as the opposing fluid pressure is relieved, the spring tension forces the check ball back onto its seat.

A *poppet valve* can be a ball or a flat disc. In either case, the poppet valve acts to block fluid flow. Often the poppet valve has a stem to guide the valve's operation. The stem normally fits into a hole acting as a guide to the valve's opening and closing. Poppet valves tend to pop open and closed, hence their name.

The most commonly used valve in a valve body is the *spool valve*. The lands of a spool valve are large diameter surfaces that fit into a bore. There is a minimum of two lands per valve, and a stem connects the lands. Between the two lands is the valve's valley, which forms a fluid pressure chamber between the lands and valve body bore. Fluid flow can be directed into other passages, depending on the spool valve and valve body design.

The land rides on a very thin film of fluid in the valve body bore. The land must be treated very carefully because any damage, even a small score or scratch, can impair smooth valve operation. As the spool valve moves, the land covers (closes) or uncovers (opens) ports in the valve body. When a spool valve cycles back and forth between the open and the exhaust positions, it is called a regulating valve.

Pressures

Line pressure is also referred to as mainline pressure. Line pressure is the hydraulic pressure that operates apply devices and is the source for all other pressures in the transmission. It is developed by the transmission's pump and is regulated by the pressure regulator.

Converter pressure transmits torque and keeps transmission fluid circulating into and out of the torque converter, which reduces the formation of air bubbles and aids cooling.

There are times in the automatic transmission's operation when fluid pressure must be increased above its baseline pressure. This increase is needed to hold bands and clutches more tightly and to raise the point at which shifting takes place. Increasing pressure above normal line pressures allows operational load flexibility, which is needed, for example, when towing trailers.

There are two methods used to monitor vehicle and engine load: the TP sensor and the MAP sensor. The PCM can then anticipate the load on the engine and the driver's demands and shift the transmission accordingly. Older transmissions used a vacuum modulator and/or a shift cable to accomplish this task.

Most vehicles use an electronically operated, PCM-controlled pressure control solenoid to control the line pressure. The PCM cycles the solenoid to produce the desired line pressure according to engine load and vehicle demands.

The manual valve is a spool valve operated manually by the driver and gear selector linkage. When the driver selects a gear position, the gear selector linkage positions the manual valve in the valve body. The manual valve directs line pressure to the correct combination of circuits, which produces the range desired for the driver's requirements.

Shift Feel

All transmissions are designed to change gears at the correct time, according to engine speed, load, and driver intent. However, transmissions are also designed to provide for a positive change of gear ratios without jarring the driver or passengers. If a band or clutch is applied too quickly, a harsh shift will occur.

Shift feel is controlled by the pressure at which each reaction member is applied or released, the rate at which each is pressurized or exhausted, and the relative timing of the application and release of the members.

To improve shift feel during gear changes, a band is often released while a multiple-disc clutch is being applied. The timing of these two actions must be just right or both components will be released or applied at the same time, which would cause engine flare-up or driveline shudder.

Accumulators

Shift quality with a band or clutch depends on how quickly the apply device is engaged by hydraulic pressure and the amount of pressure exerted on the piston. Some apply circuits utilize an accumulator to slow down application rates without decreasing the holding force of the apply device.

An accumulator is similar to a servo in that it consists of a piston and cylinder. It works like a shock absorber and cushions the application of servos and clutches. An accumulator cushions sudden increases in hydraulic pressure by temporarily diverting some of the apply fluid into a parallel circuit or chamber. This allows the pressure to increase gradually and provides for smooth engagement of a band or clutch.

Electronic Circuits

A typical electronic control system is made up of sensors, actuators, and related wiring that is tied into a central processor called a microprocessor or computer. Most input sensors are designed to produce a voltage signal that varies within a given range (from high to low, including all points in between). A signal of this type is called an analog signal. Unfortunately, the computer doesn't understand analog signals. It can only read a digital binary signal, which is a signal that has only two values—on or off.

To overcome this communication problem, all analog voltage signals are converted to a digital format by an analog-to-digital converter (A/D converter). Some sensors like the Hall-effect switch produce a digital or square wave signal that can go directly to the microcomputer as input. The term *square wave* is used to describe the appearance of a digital circuit after it has been plotted on a graph. The abrupt changes in circuit condition (on and off) result in a series of horizontal and vertical lines that connect to form a square-shaped pattern.

In addition to A/D conversion, some voltage signals require amplification before they can be relayed to the computer. To perform this task, an input conditioner known as an amplifier is used to strengthen weak voltage signals.

After input has been generated, conditioned, and passed along to the microcomputer, it is ready to be processed for the purposes of performing work and displaying information. The portion of the microcomputer that receives sensor input and handles all calculations (makes decisions) is called the microprocessor. In order for the microprocessor to make the most informed decisions regarding system operation, sensor input is supplemented by the memory.

A computer's memory holds the programs and other data, such as vehicle calibrations, which the microprocessor refers to in performing calculations. To the computer, the program is a set of instructions or procedures that it must follow. Included in the program is information that tells the microprocessor when to retrieve input (based on temperature, time, etc.), how to process the input, and what to do with it once it has been processed.

The microprocessor works with memory in two ways: it can read information from memory or change information in memory by writing in or storing new information. During processing, the computer often receives more data than it can immediately handle. In these instances, some information is temporarily stored or written into memory until the microprocessor needs it.

Vehicle Network

The vehicle system includes a network of computers and sensors that control virtually every aspect of vehicle operation. Many vehicles have either a Powertrain Control Module (PCM), or a separate engine control module (ECM) and a transmission control module (TCM). The TCM and ECM share information from the engine and transmission sensors. The computers can be reprogrammed as necessary to fit a specific vehicle. The reprogramming capability also makes updates to the transmission software relatively simple, should an update bulletin be released from the manufacturer.

The transmission shifting, as well as the line pressure, is controlled by outputs from the computer. These outputs are dependent on inputs from various sensors from both the engine and transmission.

Sensors

All sensors perform the same basic function. They detect a mechanical condition (movement or position), chemical state, or temperature condition and change it into an electrical signal that can be used by the computer to make decisions. The computer makes decisions based on information it receives from sensors and the programmed instructions it has in its memory. Each sensor used in a particular system has a specific job to do (for example, monitor throttle position, vehicle speed, manifold pressure). Together these sensors provide enough information to help the computer form a complete picture of vehicle operation.

Reference voltage (Vref) sensors provide input to the computer by modifying or controlling a constant, predetermined voltage signal. This signal, which can have a reference value from 5 to 12 volts, is generated and sent out to each sensor by a reference voltage regulator located inside the processor. Because the computer knows that a certain voltage value has been sent out, it can indirectly interpret things like motion, temperature, and component position, based on what comes back. Variable resistors are typically used as voltage-reference sensors. The throttle position (TP) sensor tells the computer network the desired angle of the throttle and how fast the pedal has been depressed. This input is used to determine the load on the engine, and the shift points necessary to satisfy driver demand. The transmission may shift later or downshift into a lower gear on hard acceleration, or shift early on minimum acceleration. The MAP sensor tells the computer the engine load based on manifold pressure at the intake manifold. The mass air flow (MAF) is used to determine the amount of air entering the engine, as well as the density of the air. This information is also used to determine engine load.

Two other commonly used reference voltage sensors are switches and thermistors. Switches tell the computer when something is turned on or off and when a particular condition exists. Switches don't provide a variable signal; there is either a signal from them, or there is no signal. Switches are used in the transmission to tell the computer which gear position the driver desires: Park, Neutral, Drive, etc.

Pressure switches can also be used by the transmission to determine when the transmission actually shifts. The switch detects transmission fluid pressure when a particular gear is applied by the transmission. This tells the computer that a commanded shift was executed by the transmission.

Thermistors are temperature sensitive and send a varying signal to the computer based on the temperature they are subject to. In the transmission, the thermistor relays the temperature of the transmission fluid. The computer can alter the shifting of the transmission.

Voltage generating sensors include components like the Hall-effect switch, oxygen sensor (zirconium dioxide), and knock sensor (piezoelectric), which are capable of producing their own input voltage signal. This varying voltage signal, when received by the computer, enables the computer to monitor and adjust for changes in the operation of various systems of the automobile. Examples of voltage generating sensors in use on the transmission include the vehicle speed sensor VSS and the input shaft speed sensor. The VSS is responsible for the vehicle speed sensor reading. The input shaft speed sensor (not used on all vehicles) is used by the computer network to determine the speed of the input shaft. The computer network can then calculate the effective gear ratio of the transmission. This allows for diagnosis of transmission shafting problems. If an invalid gear ratio is detected, a trouble code is set. Some vehicles use the engine RPM signal in lieu of an input shaft speed sensor.

Actuators

After the computer has assimilated the information and the tools used by it to process this information, it sends output signals to control devices called actuators. These actuators are solenoids, switchers, relays, or motors, which physically act or carry out a decision the computer has made.

Actuators are electromechanical devices that convert an electrical current into mechanical action. This mechanical action can then be used to open and close valves, control vacuum to other components, or open and close switches. When the computer receives an input signal indicating a change in one or more of the operating conditions, the computer determines the best strategy for handling the conditions. The computer then controls a set of actuators to achieve a desired effect or strategy goal. In order for the computer to control an actuator, it must rely on a component called an output driver.

Output drivers are located in the processor and operate by the digital commands issued by the computer. Basically, the output driver is nothing more than an electronic on/off switch that the computer uses to control the ground circuit of a specific actuator.

Shift solenoids are controlled by computer outputs so as to control the shifting of the transmission according to load and vehicle speed. Often, a combination of more than one shift solenoid is used to accomplish a transmission shift.

Pressure Control Solenoids

The Pressure control solenoid is often a pulse width modulated (PWM) controlled solenoid. Pressure is controlled by varying the amount of time the solenoid is on or off. The time that the solenoid is off, full pressure is produced, and the time the solenoid is on, a minimum pressure is produced. The time the solenoid is on is called the duty cycle. The computer is capable of turning the solenoid on and off several times a second. If more pressure is desired, the computer shortens the duty cycle, if less pressure is desired, the duty cycle is lengthened. In this manner, the line pressure can be very precisely controlled by the PCM or TCM. The duty cycle allows the computer to have a digital output (on or off) and achieve a varying output.

For actuators that cannot be controlled by a solenoid, relay, switches, or motors, the computer must turn its digitally coded instructions back into an analog format via a digital-to-analog converter.

Multiplexing

Multiplexing is an in-vehicle networking system used to transfer data between electronic modules through a serial data bus. Serial data is electronically coded information that is transmitted by one computer and received and displayed by another computer. Serial data is information that is digitally coded and transmitted in a series of data bits. The data transmission rate is referred to as the baud rate. Baud rate refers to the number of data bits that can be transmitted in a second.

With multiplexing, fewer dedicated wires are required for each function, and this reduces the size of the wiring harness. Using a serial data bus reduces the number of wires by combining the signals on a single wire through time division multiplexing. Information is sent to individual control modules that control each function, such as anti-lock braking, turn signals, power windows, and instrument panel and transmission operation.

Multiplexing also eliminates the need for redundant sensors because the data from one sensor is available to all electronic modules. Multiplexing also allows for greater vehicle content flexibility because functions can be added through software changes, rather than adding another module or modifying an existing one.

The common multiplex system is called the CAN or Controller Area Network. It is used to interconnect a network of electronic control modules.

Electronic Control of Torque Converter Clutch Engagement

Late-model automatic transmissions and transaxles are electronically controlled. Shifting and torque converter clutch engagement are regulated by a computer with programmed logic and in response to input sensors and switches. The electronic system controls these two operations by means of solenoid-operated valves. With electronic control, information about the engine, fuel, ignition, vacuum, and operating temperature is used to ensure that shifting and clutch engagement take place at exactly the right time.

A computer is used to control the converter clutch solenoid. The computer turns on the converter clutch solenoid to move a valve that allows fluid pressure to engage the converter clutch. When the computer de-energizes the converter clutch solenoid, the converter clutch disengages.

Electronically Controlled Shifting

Electronically controlled transmissions function in the same way as hydraulically based transmissions, except that a computer determines their shift points. The computer uses inputs from several different sensors and matches this information to a predetermined schedule.

The computer determines the time and condition of gear changes according to the input signals it receives and the shift schedules it has stored in its memory. Shift schedule logic

chooses the proper shift schedule for the current conditions of the transmission. It uses the shift schedule to select the appropriate gear and then determines the correct shift schedule or pattern that should be followed.

Electronically controlled transmissions rely on pressure differentials at the sides of a shift valve to hold or change a gear. The pressure differential is caused by the action of shift solenoids, which allow for changes in pressure on the side of a shift valve. The computer controls these solenoids. The solenoids do not directly control the transmission's clutches and bands. These are engaged or disengaged hydraulically. The solenoids simply control the fluid pressures in the transmission and do not perform a mechanical function.

Most electronically controlled systems are complete computer systems. There is a central processing unit, inputs, and outputs. Often the central processing unit is a separate computer designated for transmission control. This computer is often called the transmission control module (TCM). Other transmission control systems use the powertrain control module (PCM) or the body control module (BCM) to control shifting. When transmission control is not handled by the PCM, the controlling unit communicates with the PCM.

The inputs include transmission operation monitors plus some of the sensors used by the PCM. Input sensors, such as the throttle position (TP) sensor, supply information for many different systems, and the control modules share this information. The system outputs are solenoids. The transmission control systems used by various vehicle manufacturers differ mostly in the number and type of solenoids used.

Electronically shifted transmissions have systems that allow the TCM to change transmission behavior in response to operating conditions and the habits of the driver. The system monitors the condition of the engine and compensates for any changes in the engine's or transmission's performance. It also monitors and memorizes the typical driving style of the driver and the operating conditions of the vehicle. With this information, the TCM adjusts the timing of shifts and converter lockup to provide good shifting at the appropriate time. Computer systems with this capability are said to have adaptive learning capabilities.

Hybrid Systems

A hybrid electric vehicle (HEV) uses one or more electric motors and an engine to propel the vehicle. Depending on the design of the system, the engine may move the vehicle by itself, assist the electric motor while it is moving the vehicle, or it may drive a generator to charge the vehicle's batteries. The electric motor may power the vehicle by itself or assist the engine while it is propelling the vehicle. Many hybrids rely exclusively on the electric motor(s) during slow speed operation, the engine at higher speeds, and both during some certain driving conditions. Complex electronic controls monitor the operation of the vehicle, based on the current operating conditions, electronics control the engine, electric motor, and generator.

A hybrid's electric motor is powered by high-voltage batteries, which are recharged by a generator driven by the engine and through regenerative braking. Regenerative braking is the process by which a vehicle's kinetic energy can be captured while it is decelerating and braking. The electric drive motors become generators driven by the vehicle's wheels. These generators take the kinetic energy, or the energy of the moving vehicle, and changes it into energy that charges the batteries. The magnetic forces inside the generator cause the drive wheels to slow down. A conventional brake system brings the vehicle to a safe stop.

The engines used in hybrids are specially designed for the vehicle and electric assist. Therefore, they can operate more efficiently; resulting in very good fuel economy and very low tailpipe emissions. HEVs can provide the same performance, if not better, as a comparable vehicle equipped with a larger engine.

There are primarily two types of hybrids: the parallel and the series designs. A parallel hybrid electric vehicle uses either the electric motor or the gas engine to propel the vehicle, or both. The engine in a true series hybrid electric vehicle is used only to drive the generator that keeps the batteries charged. The vehicle is powered only by the electric motor(s). Most current HEVs are considered as having a series/parallel configuration because they have the features of both designs.

Although most current hybrids are focused on fuel economy, the same ideas can be used to create high-performance vehicles. Hybrid technology is also influencing off-the-road performance. By using individual motors at the front and rear drive axles, additional power can be applied to certain drive wheels, when needed.

SAFETY

In an automotive repair shop, there is great potential for serious accidents simply because of the nature of the business and the equipment used. Through carelessness, the automotive repair industry can be one of the most dangerous occupations, but the chances of being injured while working on a car are close to nil if you learn to work safely and use common sense. Shop safety is the responsibility of everyone in the shop.

Personal Protection

Some procedures, such as grinding, result in tiny particles of metal and dust being thrown off at very high speeds. These metal and dirt particles can easily get into your eyes, causing scratches or cuts on your eyeball. Pressurized gases and liquids escaping a ruptured hose or hose fitting can spray a great distance. If these chemicals get into your eyes, they can cause blindness. Dirt and sharp bits of corroded metal can easily fall into your eyes while you are working under a vehicle.

Eye protection should be worn whenever you are exposed to these risks. To be safe, you should wear safety glasses whenever you are working in the shop. Some procedures may require that you wear other eye protection in addition to safety glasses. When cleaning parts with a pressurized spray, for example, you should wear a face shield. The face shield not only gives added protection to your eyes, it also protects the rest of your face.

If chemicals such as battery acid, fuel, or solvents get into your eyes, flush them continuously with clean water. Have someone call a doctor, and get medical help immediately.

Your clothing should be well fitted and comfortable but made with strong material. Loose, baggy clothing can easily get caught in moving parts and machinery. Some technicians prefer to wear coveralls or shop coats to protect their personal clothing. Your work clothing should offer you some protection but should not restrict your movement.

Long hair and loose, hanging jewelry can create the same type of hazard as loose-fitting clothing—they can get caught in moving engine parts and machinery. If you have long hair, tie it back or tuck it under a cap.

Never wear rings, watches, bracelets, and neck chains. These can easily get caught in moving parts and cause serious injury.

Always wear leather or similar material shoes or boots with nonslip soles. Steel-tipped safety shoes can give added protection to your feet. Jogging or basketball shoes, street shoes, and sandals are inappropriate in the shop.

Good hand protection is often overlooked. A scrape, cut, or burn can limit your effectiveness at work for many days. A well-fitted pair of heavy work gloves should be worn during operations such as grinding and welding or when handling high-temperature components. Always wear approved rubber gloves when handling strong and dangerous caustic chemicals.

Many technicians wear thin, surgical-type latex gloves whenever they are working on vehicles. These offer little protection against cuts but do offer protection against disease and grease buildup under and around your fingernails. These gloves are comfortable and are quite inexpensive.

Accidents can be prevented simply by the way you act. Following are some guidelines for working in a shop. This list does not include everything you should or shouldn't do; it merely provides some things to think about.

- Never smoke while working on a vehicle or while working with any machine in the shop.

- Playing around is not fun when it sends someone to the hospital.

- To prevent serious burns, keep your skin away from hot metal parts, such as the radiator, exhaust manifold, tailpipe, catalytic converter, and muffler.

- Always disconnect electric engine cooling fans when working around the radiator. Many of these turn on without warning and can easily chop off a finger or hand. Make sure you reconnect the fan after you have completed your repairs.

- When working with a hydraulic press, make sure the pressure is applied in a safe manner. It is generally wise to stand to the side when operating the press.

- Properly store all parts and tools by putting them away in a place where people will not trip over them. This practice not only cuts down on injuries, it also reduces time wasted looking for a misplaced part or tool.

Work Area Safety

Your entire work area should be kept clean and dry. Any oil, coolant, or grease on the floor can make it slippery. To clean up oil, use commercial oil absorbent. Keep all water off the floor. Water is slippery on smooth floors, and electricity flows well through water. Aisles and walkways should be kept clean and wide enough to easily move through. Make sure the work areas around machines are large enough to operate the machine safely.

Gasoline is a highly flammable volatile liquid. Something that is *flammable* catches fire and burns easily. A *volatile* liquid is one that vaporizes very quickly. *Flammable volatile liquids* are potential firebombs. Always keep gasoline or diesel fuel in an approved safety can, and never use gasoline to clean your hands or tools.

Handle all solvents (or any liquids) with care to avoid spillage. Keep all solvent containers closed, except when pouring. Proper ventilation is very important in areas where volatile solvents and chemicals are used. Solvents and other combustible materials must be stored in approved and designated storage cabinets or rooms with adequate ventilation. Never light matches or smoke near flammable solvents and chemicals, including battery acids.

Oily rags should also be stored in an approved metal container. When oily, greasy, or paint-soaked rags are left lying about or are not stored properly, they can spontaneously combust. Spontaneous combustion refers to a fire that starts by itself, without a match.

Disconnecting the vehicle's battery before working on the electrical system or before welding can prevent fires caused by a vehicle's electrical system. To disconnect the battery, remove the negative or ground cable from the battery and position it away from the battery.

Know where all the shop's fire extinguishers are located. Fire extinguishers are clearly labeled as to their type and the types of fire they should be used on. Make sure you use the correct type of extinguisher for the type of fire you are dealing with. A multipurpose dry chemical fire extinguisher puts out ordinary combustibles, flammable liquids, and electrical fires. Never put water on a gasoline fire—water just spreads the fire. The proper fire extinguisher smothers the flames.

During a fire, never open doors or windows unless it is absolutely necessary; the extra draft only makes the fire worse. Make sure the fire department is contacted before or during your attempt to extinguish a fire.

Tool and Equipment Safety

Careless use of simple hand tools, such as wrenches, screwdrivers, and hammers, causes many shop accidents that could be prevented. Keep all hand tools free of grease and in good condition. Tools that slip can cause cuts and bruises. If a tool slips and falls into a moving part, it can fly out and cause serious injury.

Use the proper tool for the job. Make sure the tool is of professional quality. Using poorly made tools or the wrong tools can damage parts, the tool itself, or you. Never use broken or damaged tools.

Safety around power tools is very important. Serious injury can result from carelessness. Always wear safety glasses when using power tools. If the tool is electrically powered, make sure it is properly grounded. Before using it, check the wiring for cracks in the insulation, as well as for bare wires. Also, when using electrical power tools, never stand on a wet or damp floor. Never leave a running power tool unattended.

Tools that use compressed air are called pneumatic tools. Compressed air is used to inflate tires, apply paint, and drive tools. Compressed air can be dangerous when it is not used properly.

When using compressed air, wear safety glasses or a face shield, or both. Particles of dirt and pieces of metal blown by the high-pressure air can penetrate your skin or get into your eyes.

Before using a compressed air tool, check all hose connections. Always hold an air nozzle or air control device securely when starting or shutting off the compressed air. A loose nozzle can whip suddenly and cause serious injury. Never point an air nozzle at anyone. Never use compressed air to blow dirt from your clothes or hair. Never use compressed air to clean the floor or workbench.

Always be careful when raising a vehicle on a lift or a hoist. Adapters and hoist plates must be positioned correctly to prevent damage to the underbody of the vehicle. There are specific lift points that allow the weight of the vehicle to be evenly supported by the adapters or hoist plates. The correct lift points can be found in the vehicle's service manual. Before operating any lift or hoist, carefully read the operating manual and follow the operating instructions.

Once you feel the lift supports are properly positioned under the vehicle, raise the lift until the supports contact the vehicle. Then, check the supports to make sure they are in full contact with the vehicle. Shake the vehicle to make sure it is securely balanced on the lift, and then raise the lift to the desired working height. Before working under a car, make sure the lift's locking devices are engaged.

A vehicle can be raised off the ground by a hydraulic jack. The jack's lifting pad must be positioned under an area of the vehicle's frame or at one of the manufacturer's recommended lift points. Never place the pad under the floor pan or under steering and suspension components, which are easily damaged by the weight of the vehicle. Always position the jack so the wheels of the vehicle can roll as it is being raised.

Safety stands, also called jack stands, should be placed under a sturdy chassis member, such as the frame or axle housing, to support the vehicle after it has been raised by a jack. Once the safety stands are in position, the hydraulic pressure in the jack should be slowly released until the weight of the vehicle is on the stands. Never move under a vehicle when it is supported only by a hydraulic jack. Rest the vehicle on the safety stands before moving under the vehicle.

Heavy parts of the automobile, such as engines, are removed with chain hoists or cranes. Cranes often are called cherry pickers. To prevent serious injury, chain hoists and cranes must be properly attached to the parts being lifted. Always use bolts with enough strength to support the object being lifted. After you have attached the lifting chain or cable to the part that is being removed, have your instructor check it. Place the chain hoist or crane directly over the assembly, then attach the chain or cable to the hoist.

Parts cleaning is a necessary step in most repair procedures. Always wear the appropriate protection when using chemical, abrasive, and thermal cleaners.

Air Bag Safety

When service is performed on any air bag system component, always disconnect the negative battery cable, isolate the cable end, and wait for the amount of time specified by the vehicle manufacturer before proceeding with the necessary diagnosis or service. The average waiting period is two minutes, but some vehicle manufacturers specify up to ten minutes. Failure to observe this precaution may cause accidental air bag deployment and personal injury.

Replacement air bag system parts must have the same part number as the original part. Replacement parts of lesser or questionable quality must not be used. Improper or inferior components may result in inappropriate air bag deployment and injury to the vehicle occupants.

Do not strike or jar a sensor or an air bag system diagnostic monitor (ASDM). This may cause air bag deployment or make the sensor inoperative. Accidental air bag deployment may cause personal injury, and an inoperative sensor may result in air bag deployment failure, causing personal injury to vehicle occupants.

All sensors and mounting brackets must be properly torqued to ensure correct sensor operation before an air bag system is powered up. If sensor fasteners do not have the proper torque, improper air bag deployment may result in injury to vehicle occupants.

When working on the electrical system on an air-bag-equipped vehicle, use only the vehicle manufacturer's recommended tools and service procedures. The use of improper tools or service procedures may cause accidental air bag deployment and personal injury. For example, do not use 12V or self-powered test lights when servicing the electrical system on an air-bag-equipped vehicle.

Working Safely on High-Voltage Systems

Electric drive vehicles (battery-operated, hybrid, and fuel cell electric vehicles) have high-voltage electrical systems (from 42 volts to 650 volts). These high voltages can kill you! Fortunately, most high-voltage circuits are identifiable by size and color. The cables have thicker insulation and are typically colored orange. The connectors are also colored orange. On some vehicles, the high-voltage cables are enclosed in an orange shielding or casing, again the orange indicates high voltage. In addition, the high-voltage battery pack and most high-voltage components have "High Voltage" caution labels. Be careful not to touch these wires and parts.

Wear insulating gloves, commonly called "lineman's gloves," when working on or around the high-voltage system. These gloves must be class "0" rubber insulating gloves, rated at 1000-volts. Also, to protect the integrity of the insulating gloves, as well as you, wear leather gloves over the insulating gloves while doing a service.

Make sure they have no tears, holes or cracks and that they are dry. Electrons are very small and can enter through the smallest of holes in your gloves. The integrity of the gloves should be checked before using them. To check the condition of the gloves, blow enough air into each one so they balloon out. Then fold the open end over to seal the air in. Continue to slowly fold that end of the glove toward the fingers. This will compress the air. If the glove continues to balloon as the air is compressed, it has no leaks. If any air leaks out, the glove should be discarded. All gloves, new and old, should be checked before they are used.

There are other safety precautions that should always be adhered to when working on an electric drive vehicle:

- Always adhere to the safety guidelines given by the vehicle's manufacture.
- Obtain the necessary training before working on these vehicles.
- Be sure to perform each repair operation following the test procedures defined by the manufacturer.
- Disable or disconnect the high-voltage system before performing services to those systems. Do this according to the procedures given by the manufacturer.
- Anytime the engine is running in a hybrid vehicle, the generator is producing high voltage and care must be taken to prevent being shocked.
- Before doing any service to an electric drive vehicle, make sure the power to the electric motor is disconnected or disabled.
- Systems may have a high-voltage capacitor that must be discharged after the high-voltage system has been isolated. Make sure to wait the prescribed amount of time (normally about 10 minutes) before working on or around the high-voltage system.

- After removing a high-voltage cable, cover the terminal with vinyl electrical tape.
- Always use insulated tools.
- Alert other technicians that you are working on the high-voltage systems with a warning sign such as "High-Voltage Work: Do Not Touch!"
- Always install the correct type of circuit protection device into a high-voltage circuit.
- Many electric motors have a strong permanent magnet in them; individuals with a pacemaker should not handle these parts.
- When an electric drive vehicle needs to be towed into the shop for repairs, make sure it is not towed on its drive wheels. Doing this will drive the generator(s), which can overcharge the batteries and cause them to explode. Always tow these vehicles with the drive wheels off the ground or move them on a flat bed.

Vehicle Operation

When a customer brings in a vehicle for service, certain driving rules should be followed to ensure your safety and the safety of those working around you. For example, before moving a car into the shop, buckle your safety belt. Make sure no one is nearby, the way is clear, and there are no tools or parts under the car before you start the engine.

Check the brakes before putting the vehicle in gear. Then drive slowly and carefully in and around the shop.

If the engine must be running while you are working on the car, block the wheels to prevent the car from moving. Put the transmission in park for automatic transmissions and set the parking (emergency) brake. Never stand directly in front of or behind a running vehicle.

Run the engine only in a well-ventilated area to avoid the danger of poisonous carbon monoxide (CO) in the engine exhaust. CO is an odorless but deadly gas. Most shops have an exhaust ventilation system; always use it. Connect the hose from the vehicle's tailpipe to the intake for the vent system. Make sure the vent system is turned on before running the engine. If the work area does not have an exhaust venting system, use a hose to direct the exhaust out of the building.

Static Safeguards

Some manufacturers mark certain components and circuits with a code or symbol to warn technicians that they are sensitive to electrostatic discharge. Static electricity can destroy or render a component useless.

When handling any electronic part, especially those that are static sensitive, follow the guidelines below to reduce the possibility of electrostatic build-up on your body and the inadvertent discharge to the electronic part. If you are not sure if a part is sensitive to static, treat it as if it were.

1. Always touch a known good ground before handling the part. This should be repeated while handling the part and more frequently after sliding across a seat, sitting down from a standing position, or walking a distance.
2. Avoid touching the electrical terminals of the part, unless you are instructed to do so in the written service procedures. It is good practice to keep your fingers off all electrical terminals, as the oil from your skin can cause corrosion.
3. When you are using a voltmeter, always connect the negative meter lead first.
4. Do not remove a part from its protective package until it is time to install the part.
5. Before removing the part from its package, ground yourself and the package to a known good ground on the vehicle.

Some tool manufacturers have grounding straps that are available. These are designed to fit on you, at one end, and be fastened to a good ground on the other end. They are some of the best safeguards against static electricity.

HAZARDOUS MATERIALS AND WASTES

A typical shop contains many potential health hazards for those working in it. These hazards can cause injury, sickness, health impairments, discomfort, and even death. Here is a short list of the different classes of hazards.

- Chemical hazards are caused by high concentrations of vapors, gases, or solids in the form of dust.
- Hazardous wastes are substances that result from a service.
- Physical hazards include excessive noise, vibration, pressures, and temperatures.
- Ergonomic hazards are conditions that impede normal or proper body position and motion.

Many government agencies are charged with ensuring safe work environments for all workers, including the Occupational Safety and Health Administration (OSHA), Mine Safety and Health Administration (MSHA), and National Institute for Occupational Safety and Health (NIOSH). These agencies, in addition to state and local governments, have instituted regulations that must be understood and followed. Everyone in a shop is responsible for adhering to these regulations.

An important part of a safe work environment is the employees' knowledge of potential hazards. Right-to-know laws concerning all chemicals protect every employee in the shop. The general intent of right-to-know laws is for employers to provide their employees with a safe working place as it relates to hazardous materials.

All employees must be trained about their rights under the legislation, the nature of the hazardous chemicals in their workplace, and the contents of the labels on the chemicals. All the information about each chemical must be posted on Material Safety Data Sheets (MSDS) and must be accessible. The manufacturer of the chemical must give these sheets to its customers on request. They detail the chemical composition and precautionary information for all products that can present a health or safety hazard.

Employees must become familiar with the general uses, protective equipment, accident and spill procedures, and any other information regarding the safe handling of the hazardous material. This training must be given to employees annually and provided to new employees as part of their job orientation.

All hazardous material must be properly labeled, indicating what health, fire, or reactivity hazard it poses and what protective equipment is necessary when handling each chemical. The manufacturer of the hazardous materials must provide all warnings and precautionary information, which must be read and understood by the user before use. A list of all hazardous materials used in the shop must be posted for employees to see.

Shops must maintain documentation on the hazardous chemicals in the workplace, proof of training programs, records of accidents and spill incidents, satisfaction of employee requests for specific chemical information via the MSDS, and a general right-to-know compliance procedure.

When handling any hazardous materials or hazardous waste, make sure you follow the required procedures for handling such material. Wear the proper safety equipment listed on the MSDS, including approved respirator equipment.

Some of the common hazardous materials that automotive technicians use are cleaning chemicals, fuels (gasoline and diesel), paints and thinners, battery electrolyte (acid), used engine oil, used transmission fluid, refrigerants, and engine coolant (antifreeze).

Many repair and service procedures generate what are known as hazardous wastes. Dirty solvents and cleaners are good examples of hazardous wastes. Something is classified as a hazardous waste if it is on the Environmental Protection Agency (EPA) list of known harmful materials or has one or more of the following characteristics.

- *Ignitability.* A liquid with a flash point below 1408 F or a solid that can spontaneously ignite.
- *Corrosivity.* A substance that dissolves metals and other materials or burns the skin.
- *Reactivity.* Any material that reacts violently with water or other materials or releases cyanide gas, hydrogen sulfide gas, or similar gases when exposed to low-pH acid solutions, including material that generates toxic mists, fumes, vapors, and flammable gases.
- *EP toxicity.* Materials that leach one or more of eight heavy metals in concentrations greater than 100 times primary drinking water standard concentrations.

Complete EPA lists of hazardous wastes can be found in the Code of Federal Regulations. It should be noted that no material is considered hazardous waste until the shop is finished using it and ready to dispose of it.

The following list describes the recommended procedure for dealing with some common hazardous wastes. Always follow these and any other mandated procedures.

Fluids Collect all used fluids, such as automatic transmission fluid (ATF), and store them with similar fluids in a container marked to identify the contents. Do not mix fluids, except as allowed by the recycler. Handle used fluids in the same way as engine oil.

Engine oil Recycle oil. Set up equipment such as a drip table or screen table with a used oil collection bucket to collect oils dripping from parts. Place drip pans underneath vehicles that are leaking fluids onto the storage area. Do not mix other wastes with used oil, except as allowed by your recycler. Used oil generated by a shop (and oil received from household "do-it-yourself" generators) may be burned on site in a commercial space heater. Used oil also may be burned for energy recovery. Contact state and local authorities to determine requirements and to obtain necessary permits.

Oil filters Drain for at least 24 hours, crush, and recycle used oil filters.

Batteries Recycle batteries by sending them to a reclaimer or back to the distributor. Keep shipping receipts to demonstrate that you have recycled. Store batteries in a watertight, acid resistant container. Inspect batteries for cracks and leaks when they come in. Treat a dropped battery as if it were cracked. Acid residue is hazardous because it is corrosive and may contain lead and other toxics. Neutralize spilled acid by using baking soda or lime, and dispose of as hazardous material.

Metal residue from machining Collect metal filings when machining metal parts. Keep separate and recycle if possible. Prevent metal filings from falling into a storm sewer drain.

Refrigerants Recover or recycle refrigerants during the service and disposal of motor vehicle air conditioners and refrigeration equipment. It is not allowable to knowingly vent refrigerants to the atmosphere. Recovering or recycling during servicing must be performed by an EPA-certified technician using certified equipment and following specified procedures.

Solvents Replace hazardous chemicals with less toxic alternatives that have equal performance. For example, substitute water-based cleaning solvents for petroleum-based solvent degreasers. To reduce the amount of solvent used when cleaning parts, use a two-stage process—dirty solvent followed by fresh solvent. Hire a hazardous waste management service to clean and recycle solvents. (Some spent solvents must be disposed of as hazardous waste unless they are recycled properly.) Store solvents in closed containers to prevent evaporation. Evaporation of solvents contributes to ozone depletion and smog formation. In addition, the residue from evaporation must be treated as a hazardous waste. Properly label spent solvents and store on drip pans or in diked areas and only with compatible materials.

Containers Cap, label, cover, and properly store above ground and outdoors any liquid containers and small tanks within a diked area and on a paved impermeable surface to prevent spills from running into surface or ground water.

Other solids Store materials such as scrap metal, old machine parts, and worn tires under a roof or tarpaulin to protect them from the elements and to prevent potentially contaminated runoff. Consider recycling tires by retreading them.

Liquid recycling Collect and recycle coolants from radiators. Store transmission fluids, brake fluids, and solvents containing chlorinated hydrocarbons separately, and recycle or dispose of them properly.

Shop towels or rags Keep waste towels in a closed container marked "Contaminated Shop Towels Only." To reduce costs and liabilities associated with disposal of used towels, which can be classified as hazardous wastes, investigate using a laundry service that is able to treat the wastewater generated from cleaning the towels.

Waste storage Always keep hazardous waste separate, properly labeled, and sealed in the recommended containers. The storage area should be covered and may need to be fenced and locked if vandalism could be a problem. Select a licensed hazardous waste hauler after seeking recommendations and reviewing the firm's permits and authorizations.

TOOLS AND EQUIPMENT

Many different tools, testing and measuring equipment are used to service automatic transmissions. NATEF has identified many of these and has said that an automatic transmission technician must know what they are and how and when to use them. The tools and equipment listed by NATEF are covered in the following discussion. Also included are the tools and equipment you will use while completing the job sheets. Although you will be using common hand tools, they are not part of this discussion. You should already know what they are and how to use and care for them.

Circuit Tester

Circuit testers are used to identify short and open circuits in any electrical circuit. Low-voltage testers are used to troubleshoot 6- to 12-volt circuits. A circuit tester, commonly called a test light, looks like a stubby ice pick. Its handle is transparent and contains a light bulb. A probe extends from one end of the handle and a ground clip and wire from the other end. When the ground clip is attached to a good ground and the probe touched to a live connector, the bulb in the handle will light up. If the bulb does not light, voltage is not available at the connector.

WARNING: *Do not use a conventional 12-V test light to diagnose components and wires in electronic systems. The current draw of these test lights may damage computers and system components. High-impedance test lights are available for diagnosing electronic systems.*

A self-powered test light is called a continuity tester. It is used on non-powered circuits. It looks like a regular test light, except that it has a small internal battery. When the ground clip is attached to the negative side of a component and the probe touched to the positive side, the lamp will light if there is continuity in the circuit. If an open circuit exists, the lamp will not light. Do not use any type of test light or circuit tester to diagnose automotive air bag systems.

Voltmeter

A voltmeter has two leads: a red positive lead and a black negative lead. The red lead should be connected to the positive side of the circuit or component. The black should be connected to ground or to the negative side of the component. Voltmeters should be connected across the circuit being tested.

The voltmeter measures the voltage available at any point in an electrical system. A voltmeter can also be used to test voltage drop across an electrical circuit, component, switch, or connector. A voltmeter can also be used to check for proper circuit grounding.

Ohmmeter

An ohmmeter measures the resistance to current flow in a circuit. In contrast to the voltmeter, which uses the voltage available in the circuit, the ohmmeter is battery powered. The circuit being tested must be open. If the power is on in the circuit, the ohmmeter will be damaged.

The two leads of the ohmmeter are placed across or in parallel with the circuit or component being tested. The red lead is placed on the positive side of the circuit and the black lead is placed on the negative side of the circuit. The meter sends current through the component and determines the amount of resistance based on the voltage dropped across the load. The scale of an ohmmeter reads from zero to infinity. A zero reading means there is no resistance in the circuit and may indicate a short in a component that should show a specific resistance. An infinite reading indicates a number higher than the meter can measure. This usually is an indication of an open circuit.

Ohmmeters are also used to trace and check wires or cables. Assume that one wire of a four-wire cable is to be found. Connect one probe of the ohmmeter to the known wire at one end of the cable and touch the other probe to each wire at the other end of the cable. Any evidence of resistance, such as meter needle deflection, indicates the correct wire. Using this same method, you can check a suspected defective wire. If resistance is shown on the meter, the wire is sound. If no resistance is measured, the wire is defective (open). If the wire is okay, continue checking by connecting the probe to other leads. Any indication of resistance indicates that the wire is shorted to one of the other wires and that the harness is defective.

Ammeter

An ammeter measures current flow in a circuit. The ammeter must be placed into the circuit or in series with the circuit being tested. Normally, this requires disconnecting a wire or connector from a component and connecting the ammeter between the wire or connector and the component. The red lead of the ammeter should always be connected to the side of the connector closest to the positive side of the battery and the black lead should be connected to the other side.

It is much easier to test current using an ammeter with an inductive pickup. The pickup clamps around the wire or cable being tested. These ammeters measure amperage based on the magnetic field created by the current flowing through the wire. This type of pickup eliminates the need to separate the circuit to insert the meter.

Because ammeters are built with very low internal resistance, connecting them in series does not add any appreciable resistance to the circuit. Therefore, an accurate measurement of the current flow can be taken.

DMMs

It's not necessary for a technician to own separate meters to measure volts, ohms, and amps; a multimeter can be used instead. Top-of-the-line multimeters are multifunctional. Most test volts, ohms, and amperes in both DC and AC. Usually there are several test ranges provided for each of these functions. In addition to these basic electrical tests, multimeters also test engine rpm, duty cycle, pulse width, diode condition, frequency, and even temperature. The technician selects the desired test range by turning a control knob on the front of the meter.

Multimeters are available with either analog or digital displays, but the most commonly used multimeter is the digital volt/ohmmeter (DVOM), which is often referred to as a digital multimeter (DMM). There are several drawbacks to using analog-type meters for testing electronic control systems. Many electronic components require very precise test results. Digital meters can measure volts, ohms, or amperes in tenths and hundredths. Another problem with analog meters is their low internal resistance (input impedance). The low input impedance allows too much current to flow through circuits and should not be used on delicate electronic devices.

Digital meters, on the other hand, have a high input impedance, usually at least 10 megohms (10 million ohms). Metered voltage for

resistance tests is well below 5 volts, reducing the risk of damage to sensitive components and delicate computer circuits. A high-impedance digital multimeter must be used to test the voltage of some components and systems such as an oxygen (O_2) sensor circuit. If a low-impedance analog meter is used in this type of circuit, the current flow through the meter is high enough to damage the sensor.

DMMs have either an "auto range" feature, in which the appropriate scale is automatically selected by the meter, or they must be set to a particular range. In either case, you should be familiar with the ranges and the different settings available on the meter you are using. To designate particular ranges and readings, meters display a prefix before the reading or range. If the meter has a setting for mAmps, that means the readings will be given in milli-amps or 1/1000th of an amp. Ohmmeter scales are expressed as a multiple of tens or use the prefix K or M. K stands for Kilo or 1000. A reading of 10K ohms equals 10,000 ohms. An M stands for Mega or 1,000,000. A reading of 10M ohms equals 10,000,000 ohms. When using a meter with an auto range, make sure you note the range being used by the meter. There is a big difference between 10 ohms and 10,000,000 ohms.

After the test range has been selected, the meter is connected to the circuit in the same way as if it were an individual meter.

When using the ohmmeter function, the DMM will show a zero or close to zero when there is good continuity. If the continuity is very poor, the meter will display an infinite reading. This reading is usually shown as a blinking "1.000", a blinking "1", or an "OL". Before taking any measurement, calibrate the meter. This is done by holding the two leads together and adjusting the meter reading to zero. Not all meters need to be calibrated; some digital meters automatically calibrate when a scale is selected. On meters that require calibration, it is recommended that the meter be zeroed after changing scales.

Multimeters may also have the ability to measure duty cycle, pulse width, and frequency. All of these represent voltage pulses caused by the turning on and off of a circuit or the increase and decrease of voltage in a circuit. Duty cycle is a measurement of the amount of time something is on compared to the time of one cycle and is measured in a percentage.

Pulse width is similar to duty cycle except that it is the exact time something is turned on and is measured in milliseconds. When measuring duty cycle, you are looking at the amount of time something is on during one cycle.

The number of cycles that occur in one second is called the frequency. The higher the frequency, the more cycles occur in a second. Frequencies are measured in Hertz. One Hertz is equal to one cycle per second.

Lab Scopes

An oscilloscope is a visual voltmeter. An oscilloscope converts electrical signals to a visual image representing voltage changes over a specific period of time. This information is displayed in the form of a continuous voltage line called a waveform pattern or trace.

An oscilloscope screen is a cathode ray tube (CRT), which is very similar to the picture tube in a television set. High voltage from an internal source is supplied to an electron gun in the back of the CRT when the oscilloscope is turned on. This electron gun emits a continual beam of electrons against the front of the CRT. The external leads on the oscilloscope are connected to deflection plates above and below and on each side of the electron beam. When a voltage signal is supplied from the external leads to the deflection plates, the electron beam is distorted and strikes the front of the screen in a different location to indicate the voltage signal from the external leads.

An upward movement of the voltage trace on an oscilloscope screen indicates an increase in voltage, and a downward movement of this trace represents a decrease in voltage. As the voltage trace moves across an oscilloscope screen, it represents a specific length of time.

The size and clarity of the displayed waveform is dependent on the voltage scale and the time reference selected. Most scopes are equipped with controls that allow voltage and time interval selection. It is important, when choosing the scales, to remember that a scope displays voltage over time.

Dual-trace oscilloscopes can display two different waveform patterns at the same time. This makes cause and effect analysis easier.

With a scope, precise measurement is possible. A scope will display any change in voltage as it occurs. This is especially important for diagnosing intermittent problems.

The screen of a lab scope is divided into small divisions of time and voltage. Time is represented by the horizontal movement of the waveform. Voltage is measured with the vertical position of the waveform. Since the scope displays voltage over time, the waveform moves from the left (the beginning of measured time) to the right (the end of measured time). The value of the divisions can be adjusted to improve the view of the voltage waveform.

Since a scope displays actual voltage, it will display any electrical noise or disturbances that accompany the voltage signal. Noise is primarily caused by radio frequency interference (RFI), which may come from the ignition system. RFI is an unwanted voltage signal that rides on a signal. This noise can cause intermittent problems with unpredictable results. The noise causes slight increases and decreases in the voltage. When a computer receives a voltage signal with noise, it will try to react to the minute changes. As a result, the computer responds to the noise rather than the voltage signal.

Graphing Multimeter

One of the latest trends in diagnostic tools is a graphing digital multimeter. These meters display readings over time, similar to a lab scope. The graph displays the minimum and maximum readings on a graph, as well as displaying the current reading. By observing the graph, a technician can identify any undesirable changes during the transition from a low reading to a high reading, or vice versa. These glitches are some of the more difficult problems to identify without a graphing meter or a lab scope.

Scan Tools

A scan tool is a microprocessor designed to communicate with the vehicle's computer network. Connected to the network through diagnostic connectors, a scan tool can access trouble codes, run tests to check system operations, and monitor the activity of the system. Trouble codes and test results are displayed on an LED screen, or printed out on the scanner printer.

Scan tools are capable of testing many onboard computer systems, such as climate controls, transmission controls, engine computers, antilock brake computers, air bag computers, and suspension computers, depending on the year and make of the vehicle and the type of scan tester. In many cases, the technician must select the computer system to be tested with the scanner after it has been connected to the vehicle.

The scan tool is connected to the diagnostic connector on the vehicle. There are many different scan tools available. Some are a combination of other diagnostic tools, such as a lab scope and graphing multimeter. These may have the following capabilities:

- Retrieve DTCs
- Monitor system operational data
- Reprogram the vehicle's electronic control modules
- Perform systems diagnostic tests
- Display appropriate service information, including electrical diagrams
- Display TSBs
- Troubleshooting instructions
- Easy tool updating through an Internet connection

The vehicle's computer sets trouble codes when a signal or self-test result is entirely out of its normal range. The codes help technicians identify the cause of the problem when this is the case. If a signal is within its normal range but is still not correct, the vehicle's computer will not display a trouble code. However, a problem may still exist.

Some scan tools work directly with a PC through uncabled communication links, such as Bluetooth. Others use a Personal Digital Assistant (PDA). These are small hand-held units that allow you to read diagnostic trouble codes (DTCs), monitor the activity of sensors, and inspection/maintenance system test results to quickly determine what service the vehicle requires. Most of these scan tools also have the ability to:

- Perform system and component tests
- Report test results of monitored systems
- Exchange files between a PC and PDA
- View and print files on a PC
- Print DTC/Freeze Frame
- Generate emissions reports
- IM/Mode 6 information
- Display related TSBs
- Display full diagnostic code descriptions
- Observe live sensor data
- Update the scan tool as a manufacturer's interfaces change

Computer Memory Saver

Memory savers are an external power source used to maintain the memory circuits in electronic accessories and the engine, transmission, and body computers when the vehicle's battery is disconnected. The saver is plugged into the vehicle's cigar lighter outlet. It can be powered by a 9- or 12-volt battery.

Hybrid Tools

A hybrid vehicle is an automobile and as such is subject to many of the same problems as a conventional vehicle. Most systems in a hybrid vehicle are diagnosed in the same way as well. However, a hybrid vehicle has unique systems that require special procedures and test equipment. It is imperative to have good information before attempting to diagnose these vehicles. Also, make sure you follow all test procedures precisely as they are given.

An important diagnostic tool is a DMM. However, this is not the same DMM used on a conventional vehicle. The meter used on hybrids and other electric-powered vehicles should be classified as a category III meter. There are basically four categories for low-voltage electrical meters, each built for specific purposes and to meet certain standards. Low-voltage, in this case, means voltages less than 1000-volts. The categories define how safe a meter is when measuring certain circuits. The standards for the various categories are defined by the American National Standards Institute (ANSI), the International Electrotechnical Commission (IEC), and the Canadian Standards Association (CSA). A CAT III meter is required for testing hybrid vehicles because of the high voltages, three-phase current, and the potential for high transient voltages. Transient voltages are voltage surges or spikes that occur in AC circuits. To be safe, you should have a CAT III-1000 V meter. A meter's voltage rating reflects its ability to withstand transient voltages. Therefore, a CAT III-1000 V meter offers much more protection than a CAT III meter rated at 600 volts.

Another important tool is an insulation resistance tester. These can check for voltage leakage through the insulation of the high-voltage cables. Obviously no leakage is desired and any leakage can cause a safety hazard as well as damage to the vehicle. Minor leakage can also cause hybrid system-related driveability problems. This meter is not one commonly used by automotive technicians, but should be for anyone who might service a damaged hybrid vehicle, such as doing body repair. This should also be a CAT III meter and may be capable of checking resistance and voltage of circuits like a DMM.

To measure insulation resistance, system voltage is selected at the meter and the probes placed at their test position. The meter will display the voltage it detects. Normally, resistance readings are taken with the circuit de-energized unless you are checking the effectiveness of the cable or wire insulation. In this case, the meter is measuring the insulation's effectiveness and not its resistance.

The probes for the meters should have safety ridges or finger positioners. These help prevent physical contact between your fingertips and the meter's test leads.

Vacuum Gauge

Measuring intake manifold vacuum is another way to diagnose the condition of an engine. This is important when diagnosing automatic transmissions. A transmission responds to engine load, and engine load affects engine vacuum. As load increases, engine vacuum decreases. This means that a poorly running engine will be seen by the transmission as always having a load on it, which results in poor shift quality and inefficient shift timing.

Manifold vacuum is tested with a vacuum gauge. Vacuum is formed on a piston's intake stroke. As the piston moves down, it lowers the pressure of the air in the cylinder—if the cylinder is sealed. This lower cylinder pressure is called engine vacuum. If there is a leak, atmospheric pressure forces air into the cylinder and the resultant pressure is not as low. The reason atmospheric pressure enters is simply that whenever low and high pressures exist, high pressure always moves toward the low pressure.

Vacuum is measured in inches of mercury (in./Hg) and in kiloPascals (kPa) or millimeters of mercury (mm/Hg).

To measure vacuum, a flexible hose on the vacuum gauge is connected to a source of manifold vacuum. Sometimes this requires removing a plug from the manifold and installing a special fitting.

The test is made with the engine cranking or running. A good vacuum reading is typically at least 16 in./Hg. However, a reading of 15 to 20 in./Hg (50 to 65 kPa) is normally acceptable. Since the intake stroke of each cylinder occurs at a different time, the production of vacuum occurs in pulses. If the amount of vacuum produced by each cylinder is the same, the vacuum gauge will show a steady reading. If one or more cylinders are producing different amounts of vacuum, the gauge will show a fluctuating reading.

Hydraulic Pressure Gauge Set

A common diagnostic tool for automatic transmissions is a hydraulic pressure gauge. A pressure gauge measures pressure in pounds per square inch (psi) and/or kPa. The gauge is normally part of a kit that contains various fittings and adapters.

Portable Crane

To remove and install a transmission, the engine is often moved out of the vehicle with the transmission. To remove or install an engine, a portable crane is used. A crane uses hydraulic pressure that is converted to a mechanical advantage and lifts the engine from the vehicle. To lift an engine, attach a pulling sling or chain to the engine. Some engines have eye plates for use in lifting. If they are not available, the sling must be bolted to the engine. The sling attaching bolts must be large enough to support the engine and must thread into the block at least 1.5 times the bolt diameter. Connect the crane to the chain. Raise the engine slightly and make sure the sling attachments are secure. Carefully lift the engine out of its compartment.

Lower the engine close to the floor so that the transmission and torque converter can be removed from the engine if necessary.

Transmission Jacks

Transmission jacks (Figure 2) are designed to help you while removing a transmission from under the vehicle. The weight of the transmission makes it difficult and unsafe to remove it without assistance or the use of a transmission jack. These jacks fit under the transmission and are typically equipped with hold-down chains. These chains are used to secure the transmission to the jack, and the transmission's weight rests on the jack's saddle.

Transmission jacks are available in two basic styles. One is used when the vehicle is raised by a hydraulic jack and is set on jack stands. The other style is used when the vehicle is raised on a lift.

Figure 2 A transmission jack.

Transaxle Removal and Installation Equipment

Removal and replacement of transversely mounted engines may require other tools. The engines of some FWD vehicles are removed by lifting them from the top. Others must be removed from the bottom, which requires different equipment. Make sure you follow the instructions given by the manufacturer and use the appropriate tools and equipment. The required equipment varies with manufacturer and vehicle model but most accomplish the same thing.

To remove the engine and transmission from under the vehicle, the vehicle must be raised. A crane or support fixture (or both) is used to hold the engine and transaxle assembly in place while the assembly is being readied for removal. When everything is set for removal of the assembly, the crane is used to lower the assembly onto a cradle. The cradle is similar to a hydraulic floor jack and is used to lower the assembly further so it can be rolled out from under the vehicle. The transaxle can be separated from the engine once it has been removed from the vehicle.

When the transaxle is removed as a single unit, the engine must be supported while it is in the vehicle before, during, and after transaxle removal. Special fixtures mount to the vehicle's upper frame or suspension parts (Figure 3). These

Figure 3 A transverse engine support bar.

supports have a bracket that is attached to the engine. With the bracket in place, the engine's weight is now on the support fixture, and the transmission can be removed.

Transmission/ Transaxle Holding Fixtures

Special holding fixtures should be used to support the transmission or transaxle after it has been removed from the vehicle. These holding fixtures may be standalone units or bench mounted; they allow the transmission to be easily repositioned during repair work.

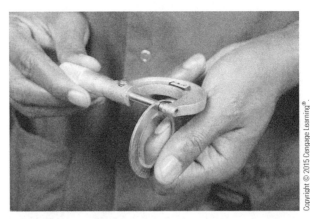

Figure 4 An outside micrometer.

Machinist's Rule

A machinist's rule is very much like an ordinary ruler. Each edge of this measuring tool is divided into increments based on a different scale. A typical machinist's rule based on the USCS measurement may have scales based on 1/8-, 1/16-, 1/32-, and 1/64-inch intervals. Of course, metric machinist rules are also available. Metric rules are usually divided into 0.5-mm and 1-mm increments.

Some machinist's rules are based on decimal intervals. These are typically divided into 1/10-, 1/100-, and 1/1,000-inch (0.1, 0.01, and 0.001) increments. Decimal machinist's rules are very helpful when measuring dimensions that are specified in decimals, so that you do not have to convert fractions to decimals.

Micrometers

A micrometer is used to measure linear outside and inside dimensions. Both outside and inside micrometers are calibrated and read in the same manner. The major components and markings of a micrometer include the frame, anvil, spindle, lock nut, sleeve, sleeve numbers, sleeve long line, thimble marks, thimble, and ratchet. Micrometers are calibrated in either inch or metric graduations and are available in a range of sizes.

To use and read a micrometer, choose the appropriate size for the object being measured (Figure 4). Typically they measure an inch; therefore, the range covered by one size of micrometer would be from 0 to 1 inch, another would measure 1 to 2 inches, and so on.

Open the jaws of the micrometer and slip the object between the spindle and anvil. While holding the object against the anvil, turn the thimble using your thumb and forefinger until the spindle contacts the object. Never clamp the micrometer tightly. Use only enough pressure on the thimble to allow the work to just fit between the anvil and spindle. To get accurate readings, you should slip the micrometer back and forth over the object until you feel a very light resistance while rocking the tool from side to side to make certain the spindle cannot be closed any further. When a satisfactory adjustment has been made, lock the micrometer. Read the measurement scale.

The graduations on the sleeve each represent 0.025 inch. To read a measurement on a micrometer, begin by counting the visible lines on the sleeve and multiplying them by 0.025. The graduations on the thimble assembly define the area between the lines on the sleeve. The number indicated on the thimble is added to the measurement shown on the sleeve. This sum is the dimension of the object.

Micrometers are available to measure in 0.0001 (ten-thousandths) of an inch. Use this type of micrometer if the specifications call for this much accuracy.

A metric micrometer is read in the same way, except that the graduations are expressed in the metric system of measurement. Each number on the sleeve represents 5 millimeters (mm) or 0.005 meter (m). Each of the 10 equal spaces between each number, with index lines alternating above and below the horizontal line, represents 0.5 mm, or five-tenths of a millimeter. Therefore, one revolution of the thimble changes the reading one space on the sleeve scale, or 0.5 mm. The beveled edge of the thimble is divided into

Copyright © 2015 Cengage Learning®.

Figure 5 A depth micrometer.

50 equal divisions with every fifth line numbered: 0, 5, 10,..., 45. Since one complete revolution of the thimble advances the spindle 0.5 mm, each graduation on the thimble is equal to one-hundredth of a millimeter. As with the inch-graduated micrometer, the separate readings are added together to obtain the total reading.

Some technicians use a digital micrometer, which is easier to read. These tools do not have the various scales; rather, the measurement is displayed and read directly off the micrometer.

Inside micrometers can be used to measure the inside diameter of a bore. To do this, place the tool inside the bore and extend the measuring surfaces until each end touches the bore's surface. If the bore is large, it might be necessary to use an extension rod to increase the micrometer's range. These extension rods come in various lengths. The inside micrometer is read in the same manner as an outside micrometer.

A depth micrometer (Figure 5) is used to measure the distance between two parallel surfaces. The sleeves, thimbles, and ratchet screws operate in the same way as other micrometers. Depth micrometers are read in the same way as other micrometers.

If a depth micrometer is used with a gauge bar, it is important to keep both the bar and the micrometer from rocking. Any movement of either part results in an inaccurate measurement.

Telescoping Gauge

Telescoping gauges are used for measuring bore diameters and other clearances. They may also be called snap gauges. They are available in sizes ranging from fractions of an inch through 6 inches. Each gauge consists of two telescoping plungers, a handle, and a lock screw. Snap gauges are normally used with an outside micrometer.

To use the telescoping gauge, insert it into the bore and loosen the lock screw. This allows the plungers to snap against the bore. Once the plungers have expanded, tighten the lock screw. Then remove the gauge and measure the expanse with a micrometer.

Small Hole Gauge

A small hole or ball gauge works just like a telescoping gauge except that it is designed for small bores. After it is placed into the bore and expanded, it is removed and measured with a micrometer. Like the telescoping gauge, the small hole gauge consists of a lock, a handle, and an expanding end. The end expands or retracts by turning the gauge handle.

Feeler Gauge

A feeler gauge is a thin strip of metal or plastic of known and closely controlled thickness. Several of these strips are often assembled together as a feeler gauge set that looks like a pocketknife. The desired thickness gauge can be pivoted away from others for convenient use. A feeler gauge set usually contains strips or leaves of 0.002- to 0.010-inch thickness (in steps of 0.001-inch) and leaves of 0.012- to 0.024-inch thickness (in steps of 0.002-inch).

A feeler gauge can be used by itself to measure piston clearance, fluid pump clearances, and other distances. It can also be used with a precision straightedge to check the flatness of a sealing surface.

Straightedge

A straightedge is no more than a flat bar machined to be totally flat and straight; to be effective, it must be flat and straight. Any surface that should be flat can be checked with a straightedge and feeler gauge set. The straightedge is placed across and at angles to the surface. At any low points on the surface, a feeler gauge can be placed between the straightedge and the surface. The size gauge that fills in the gap is the amount of warpage or distortion.

Dial Indicator

The dial indicator is calibrated in 0.001-inch (one-thousandth-inch) increments. Metric dial

Figure 6 A dial indicator with a holding fixture.

Figure 7 A pound per inch torque wrench.

indicators are also available. Both types are used to measure movement. Common uses of the dial indicator include measuring endplay (Figure 6), clutch pack clearances, and flexplate runout.

To use a dial indicator, position the indicator rod against the object to be measured. Then push the indicator toward the work until the indicator needle travels far enough around the gauge face to permit movement to be read in either direction. Zero the indicator needle on the gauge. Move the object in the direction required while observing the needle of the gauge. Always be sure the range of the dial indicator is sufficient to allow the amount of movement required by the measuring procedure. For example, never use a 1-inch indicator on a component that will move 2 inches.

Torque Wrench

Torque is the twisting force used to turn a fastener against the friction between the threads and between the head of the fastener and the surface of the component. The fact that practically every vehicle and engine manufacturer publishes a list of torque recommendations is ample proof of the importance of using proper amounts of torque when tightening nuts or bolts. The amount of torque applied to a fastener is measured with a torque-indicating or torque wrench.

A torque wrench is basically a ratchet or breaker bar with some means of displaying the amount of torque exerted on a bolt when pressure is applied to the handle. Torque wrenches are available with the various drive sizes. Sockets are inserted onto the drive and then placed over the bolt. As pressure is exerted on the bolt, the torque wrench indicates the amount of torque.

The common types of torque wrenches are available with inch-pound (Figure 7) and foot-pound increments.

- A beam torque wrench is not highly accurate. It relies on a beam metal that points to the torque reading.
- A "click"-type torque wrench clicks when the desired torque is reached. The handle is twisted to set the desired torque reading.
- A dial torque wrench has a dial that indicates the torque exerted on the wrench. The wrench may have a light or buzzer that turns on when the desired torque is reached.
- A digital readout type displays the torque and is commonly used to measure turning effort, as well as for tightening bolts. Some designs of this type torque wrench have a light or buzzer that turns on when the desired torque is reached.

Blowgun

Blowguns are used for air testing components (Figure 8) and blowing off parts during cleaning. Never point a blowgun at yourself or someone else. A blowgun snaps into one end of an air hose and directs airflow when a button is pressed. Always use an OSHA-approved air blowgun. Before using a blowgun, be sure it has not been modified to *eliminate air-bleed* holes on the side.

Gear and Bearing Pullers

Many tools are designed for a specific purpose. An example of a special tool is a gear and bearing

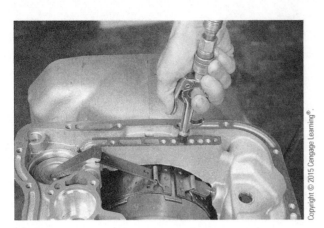

Figure 8 An air nozzle used while air testing.

puller. Many gears and bearings have a slight interference fit (press fit) when they are installed on a shaft or in a housing. Something that has a press fit has an interference fit; for example, the inside diameter of a bore is 0.001 inch smaller than the outside diameter of a shaft, so when the shaft is fitted into the bore, it must be pressed in to overcome the 0.001-inch interference fit. This press fit prevents the parts from moving on each other. These gears and bearings must be removed carefully to prevent damage to the gears, bearings, or shafts. Prying or hammering can break or bind the parts. A puller with the proper jaws and adapters should be used to remove gears and bearings. With the proper puller, the force required to remove a gear or bearing can be applied with a slight and steady motion.

Bushing and Seal Pullers and Drivers

Another commonly used group of special tools are the various designs of bushing and seal drivers (Figure 9) and pullers. Pullers are either a threaded or slide hammer type tool. Always make sure you use the correct tool for the job because bushings and seals are easily damaged if the wrong tool or procedure is used. Car manufacturers and specialty tool companies work closely together to design and manufacture special tools required to repair cars. Most of these special tools are listed in the appropriate service manuals.

Retaining Ring Pliers

An automatic transmission technician runs into many different styles and sizes of retaining rings

used to hold subassemblies together or keep them in a fixed location (Figure 10). Using the correct tool to remove and install these rings is the only safe way to work with them. All automatic transmission technicians should have an assortment of retaining ring pliers.

Special Tool Sets

Vehicle manufacturers and specialty tool companies work closely together to design and manufacture special tools required to repair transmissions. Most of these special tools are listed in the appropriate service manuals and are part of each manufacturer's Essential Tool Kit.

Service Information

Perhaps the most important tool you will use are service information. There is no way a technician can remember all the procedures and specifications needed to repair all vehicles. Good

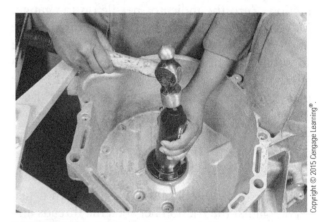

Figure 9 An example of a seal driver.

Figure 10 Snap ring pliers.

information plus knowledge allows a technician to fix a problem with the least frustration and at the lowest expense to the customer.

To obtain the correct transmission specifications and other information, you must first identify the transmission you are working on. The best source for positive identification is the Vehicle Identification Number (VIN). The transmission code can be interpreted through information given in the service information, which may also help you identify the transmission through appearance, casting numbers, or markings on the housing.

The primary source of repair and specification information for any car, van, or truck is the manufacturer. The manufacturer places service information on the Internet. The information is available for a subscription fee. Manufacturers' service information covers all repairs, adjustments, specifications, detailed diagnostic procedures, and the special tools required.

Since many technical changes occur on specific vehicles each year, manufacturers' service manuals need to be constantly updated The manufacturer provides these bulletins to dealers and repair facilities on a regular basis, over the Internet.

Service information is also available from independent companies rather than the manufacturers. However, they pay for and get most of their information from the car makers. These information contains component information, diagnostic steps, repair procedures, and specifications for several car makes Information is usually condensed and is more general in nature than the manufacturer's manuals. The condensed format allows for more coverage in less space.

Many of the larger parts manufacturers have excellent guides on the various parts they manufacture or supply. They also provide updated service bulletins on their products. Other sources for up-to-date technical information are trade magazines and trade associations.

Service information is now commonly found electronically on digital video disks (DVD) or online. Online data can be updated instantly and requires no space for physical storage. These systems are easy to use and the information is quickly accessed and displayed. The computer's keyword and mouse are used to make selections from the screen's menu. Once the information is retrieved, a technician can read it off the screen or print it out and take it to the service bay.

JOB SHEETS
REQUIRED SUPPLEMENTAL TASKS (RST)

JOB SHEET 1

Shop Safety Survey

Name _____ Station _____ Date _____

NATEF Correlation

This Job Sheet addresses the following **RST** tasks:

Shop and Personal Safety:

1. Identify general shop safety rules and procedures.

8. Identify the location and use of eyewash stations.

10. Comply with the required use of safety glasses, ear protection, gloves, and shoes during lab/shop activities

11. Identify and wear appropriate clothing for lab/shop activities

12. Secure hair and jewelry for lab/shop activities.

Objective

As a professional technician, safety should be one of your first concerns. This job sheet will increase your awareness of shop safety rules and safety equipment. As you survey your shop area and answer the following questions, you will learn how to evaluate the safeness of your workplace.

Materials

Copy of the shop rules from the instructor

PROCEDURE

Your instructor will review your progress throughout this worksheet and should sign off on the sheet when you complete it.

1. Have your instructor provide you with a copy of the shop rules and procedures.

   Have you read and understood the shop rules?   ☐ Yes ☐ No

2. Before you begin to evaluate your work area, evaluate yourself. Are you dressed to work safely? ☐ Yes ☐ No

   If no, what is wrong? _____

3. Are your safety glasses OSHA approved? ☐ Yes ☐ No

   Do they have side protection shields? ☐ Yes ☐ No

4. Look around the shop and note any area that poses a potential safety hazard.

   _____

   _____

   All true hazards should be brought to the attention of the instructor immediately.

5. What is the air line pressure in the shop? _____ psi

   What should it be? _____ psi

6. Where is the first-aid kit(s) kept in the work area?

   _____

   _____

7. Ask the instructor to show the location of, and demonstrate the use of, the eyewash station. Where is it and when should it be used?

   _____

   _____

8. What is the shop's procedure for dealing with an accident?

   _____

   _____

   _____

9. Explain how to secure hair and jewelry while working in the shop.

   _____

10. List the phone numbers that should be called in case of an emergency.

    _____

Problems Encountered

_____

_____

_____

Instructor's Comments

_____

_____

_____

JOB SHEET 2

Working in a Safe Shop Environment

Name _____ Station _____ Date _____

NATEF Correlation

This Job Sheet addresses the following **RST** tasks:

Shop and Personal Safety:

2. Utilize safe procedures for handling of tools and equipment.

3. Identify and use proper placement of floor jacks and jack stands.

4. Identify and use proper procedures for safe lift operation.

5. Utilize proper ventilation procedures for working within the lab/shop area.

6. Identify marked safety areas.

9. Identify the location of the posted evacuation routes.

Objective

This job sheet will help you work safely in the shop. Two of the basic tools a technician uses are lifts and the floor jack.

This job sheet also covers the technicians' environment.

Materials

Vehicle for hoist and jack stand demonstration
Service information

Describe the vehicle being worked on:

Year _____ Make _____ Model _____

VIN _____ Engine type and size _____

PROCEDURE

1. Are there safety areas marked around grinders and other machinery?   ☐ Yes ☐ No

2. Are the shop emergency escape routes clearly marked?   ☐ Yes ☐ No

3. Have your instructor demonstrate the exhaust gas ventilation system in the shop. Explain the importance of the ventilation system.

_____

_____

4. What types of Lifts are used in the shop?

   _____

   _____

5. Find the location of the correct lifting points for the vehicle supplied by the instructor. On the rear of this sheet, draw a simple figure showing where these lift points are.

6. Ask your instructor to demonstrate the proper use of the lift.

   Summarize the proper use of the lift.

   _____

   _____

   _____

   _____

7. Demonstrate the proper use of jack stands with the help of your instructor.

   Summarize the proper use of jack stands.

   _____

   _____

   _____

Problems Encountered

_____

_____

_____

Instructor's Comments

_____

_____

_____

JOB SHEET 3

Fire Extinguisher Care and Use

Name _____ Station _____ Date _____

NATEF Correlation

This Job Sheet addresses the following **RST** task:

Shop and Personal Safety:

7. Identify the location and the types of fire extinguishers and other fire safety equipment; demonstrate knowledge of the procedures for using fire extinguishers and other fire safety equipment.

Objective

Upon completion of this job sheet, you will be able to demonstrate knowledge of the procedures for using fire extinguishers and other fire safety equipment, and identify the location of fire extinguishers in the shop.

NOTE: *Never fight a fire that is out of control or too large. Call the fire department immediately!*

1. Identify the location of the fire extinguishers in the shop.

   _____

   _____

2. Have the fire extinguishers been inspected recently? (Look for a dated tag.)

   _____

   _____

3. What types of fires are each of the shop's fire extinguishers rated to fight?

   _____

   _____

4. What types of fires should not be used with the shop's extinguishers?

   _____

   _____

5. One way to remember the operation of a fire extinguisher is to remember the term PASS. Describe the meaning of PASS on the following lines.

   a. P

   _____

   b. A

   _____

c. S

_____

d. S

_____

Problems Encountered

_____

_____

_____

Instructor's Comments

_____

_____

_____

JOB SHEET 4

Working Safely Around Air Bags

Name _____ Station _____ Date _____

NATEF Correlation

This Job Sheet addresses the following **RST** task:

Shop and Personal Safety:

13. Demonstrate awareness of the safety aspects of supplemental restraint systems (SRS), electronic brake control systems, and hybrid vehicle high-voltage circuits.

Objective

Upon completion of this job sheet, you should be able to work safely around and with air bag systems.

Tools and Materials

A vehicle(s) with air bag

Safety glasses, goggles

Service information appropriate to vehicle(s) used

Describe the vehicle being worked on:

Year _____ Make _____ Model _____

VIN _____ Engine type and size _____

PROCEDURE

1. Locate the information about the air bag system in the service information. How are the critical parts of the system identified in the vehicle?

   _____

   _____

2. List the main components of the air bag system and describe their location.

   _____

   _____

3. There are some very important guidelines to follow when working with and around air bag systems. Look through the service information to find the answers to the questions and fill in the blanks with the correct words.

   a. Wait at least _____ minutes after disconnecting the battery before beginning any service. The reserve _____ module is capable of storing enough energy to deploy the air bag for up to _____ minutes after battery voltage is lost.

b. Never carry an air bag module by its _____ or _____, and, when carrying it, always face the trim and air bag _____ from your body. When placing a module on a bench, always face the trim and air bag _____.

c. Deployed air bags may have a powdery residue on them. _____ is produced by the deployment reaction and is converted to _____ when it comes in contact with the moisture in the atmosphere. Although it is unlikely that harmful chemicals will still be on the bag, it is wise to wear _____ and _____ when handling a deployed air bag. Immediately wash your hands after handling a deployed air bag.

d. A live air bag must be _____ before it is disposed. A deployed air bag should be disposed of in a manner consistent with the _____ and manufacturer's procedures.

e. Never use a battery- or AC-powered _____, _____, or any other type of test equipment in the system unless the manufacturer specifically says to. Never probe with a _____ for voltage.

4. Explain how an air bag sensor should be handled before it is installed on the vehicle.

Problems Encountered

_____

_____

_____

Instructor's Response

_____

_____

_____

JOB SHEET 5

High-Voltage Hazards in Today's Vehicles

Name _____ Station _____ Date _____

NATEF Correlation

This Job Sheet addresses the following **RST** task:

Shop and Personal Safety:

14. Demonstrate awareness of the safety aspects of high-voltage circuits (such as high intensity discharge (HID) lamps, ignition systems, injection systems, etc.).

Objective

Upon completion of this job sheet, you will be able to describe some of the necessary precautions to take in performing work around high-voltage hazards such as HID headlamps and ignition systems.

HID Headlamp Precautions

Describe the vehicle being worked on:

Year _____ Make _____ Model _____

VIN _____ Engine type and size _____

Materials needed

Service information for HID headlamps

Describe general operating condition:

1. List three precautions a technician should observe when working with HID headlamp systems.

   _____

   _____

   _____

2. How many volts are necessary to initiate and maintain the arc inside the bulb of this HID system?

   _____

   _____

3. A technician must never probe with a test lamp between the HID ballast and the bulb. Explain why?

   _____

High-Voltage Ignition System Precautions

4. High-voltage ignition systems can cause serious injury, especially for those who have heart problems. Name at least three precautions to take when working around high-voltage ignition systems.

_____

_____

_____

_____

Problems Encountered

_____

_____

_____

Instructor's Comments

_____

_____

_____

JOB SHEET 6

Hybrid High-Voltage and Brake System Pressure Hazards

Name _____ Station _____ Date _____

NATEF Correlation

This Job Sheet addresses the following **RST** task:

Shop and Personal Safety:

13. Demonstrate awareness of the safety aspects of supplemental restraint systems (SRS), electronic brake control systems, and hybrid vehicle high-voltage circuits.

Objective

Upon completion of this job sheet, you will be able to describe some of the necessary precautions to take while performing work around high-voltage systems such as that used in hybrid vehicles, as well as the high-pressure systems of some braking control systems.

Tools and Materials

Appropriate service information

AUTHOR'S NOTE: *According to the vehicles' manufacturers, a technician should not work on hybrid vehicles without having the specific training, which is beyond the scope of this job sheet. This job sheet assumes that the student will be accessing information only, not actually working on live high-voltage vehicles.*

Protective Gear

Goggles or safety glasses with side shields

High-voltage gloves with properly inspected liners

Orange traffic cones to warn others in the shop of a high-voltage hazard

Describe the vehicle being worked on:

Year _____ Make _____ Model _____

VIN _____ Engine type and size _____

Describe general operating condition:

_____

PROCEDURE

High Pressure Braking Systems:

NOTE: *Many vehicles have a high pressure accumulator or a high pressure pump in their braking systems. Opening one of these systems can be hazardous due to the high pressures involved.*

1. Research the vehicle you have been assigned and describe the procedure that must be followed BEFORE opening the hydraulic braking system.

_____

_____

_____

Hybrid Vehicles:

1. Special gloves with liners are to be inspected before each use when working on a hybrid vehicle.

   A. How should the integrity of the gloves be checked?

   _____

   _____

   B. How often should the gloves be tested (at a lab) for recertification?

   _____

   C. What precautions must be taken when storing the gloves?

   _____

   _____

   _____

   D. When must the gloves be worn?

   _____

   _____

   _____

2. What color are the high-voltage cables on hybrid vehicles?

   _____

3. What must be done BEFORE disconnecting the main voltage supply cable?

   _____

   _____

   _____

4. Describe the safety basis of the "one hand rule."

_____

_____

_____

5. Explain the procedure to disable high voltage on the vehicle you selected.

_____

_____

_____

6. Explain the procedure to test for high voltage to ensure that the main voltage is disconnected.

_____

_____

Problems Encountered

_____

_____

_____

Instructor's Comments

_____

_____

_____

Problems Encountered

Instructor's Comments

JOB SHEET 7

Material Data Safety Sheet Usage

Name _____ Station _____ Date _____

NATEF Correlation

This Job Sheet addresses the following **RST** task:

Shop and Personal Safety:

12. Locate and demonstrate knowledge of material safety data sheets (MSDS).

Objective

On completion of this job sheet, the student will be able to locate the MSDS folder and describe the use of an MSDS sheet on the job site.

Materials

Selection of chemicals from the shop

MSDS sheets

1. Locate the MSDS folder in the shop. It should be in a prominent location. Did you have any problems finding the folder?

   _____

   _____

2. Pick a common chemical from your tool room such as brake cleaner. Locate the chemical in the MSDS folder. What chemical did you choose?

   _____

   _____

3. What is the flash point of the chemical? _____

4. Briefly describe why the flash point is important.

   _____

   _____

5. What is the first aid if the chemical is ingested?

   _____

   _____

6. Can this chemical be absorbed through the skin?

   _____

   _____

7. What are the signs of exposure to the chemical you selected?

    _____

    _____

8. What are the primary health hazards of the chemical?

    _____

    _____

9. What is the first aid procedure for exposure to this chemical?

    _____

    _____

10. What are the recommendations for protective clothing?

    _____

    _____

Problems Encountered

_____

_____

_____

Instructor's Comments

_____

_____

_____

JOB SHEET 8

Measuring Tools and Equipment Use

Name _____ Station _____ Date _____

NATEF Correlation

This Job Sheet addresses the following **RST** tasks:

Tools and Equipment:

1. Identify tools and their usage in automotive applications.

2. Identify standard and metric designations.

3. Demonstrate safe handling and use of appropriate tools.

4. Demonstrate proper cleaning, storage, and maintenance of tools and equipment.

5. Demonstrate proper use of precision measuring tools (i.e., micrometer, dial-indicator, dial caliper).

Objective

Upon completion of this job sheet, you will be able to make measurements using micrometers, dial indicators, pressure gauges, and other measuring tools. You will also be able to demonstrate the safe handling and use of appropriate tools, and demonstrate their proper use.

Tools and Materials

Items to measure selected by the instructor

Precision measuring tools: micrometer (digital or manual), dial caliper, vacuum or pressure gauge (selected by the instructor)

PROCEDURE

Have your instructor demonstrate the measuring tools that are available to you in your shop. The tools should include both standard and metric tools.

1. Describe the measuring tools you will be using.

   _____

   _____

2. Describe the items that you will be measuring.

   _____

   _____

3. Describe any special handling or safety procedures that are necessary when using the tools.

   _____

   _____

4. Describe any special cleaning or storage that these tools might require.

   _____

   _____

5. Describe the metric unit of measurement of the tools, such as millimeters, centimeters, kilopascals, etc.

   _____

   _____

6. Describe the standard unit of measure of the tools, such as inches, pounds per square inch, etc.

   _____

   _____

7. Measure the components, list them, and record your measurements in the following table.

Item measured	Measurement taken	Unit of measure

8. Clean and store the tools. Describe the process next.

Problems Encountered

_____

_____

_____

Instructor's Comments

_____

_____

_____

JOB SHEET 9

Preparing the Vehicle for Service and Customer

Name _____ Station _____ Date _____

NATEF Correlation

This Job Sheet addresses the following **RST** tasks:

Preparing Vehicle for Service:

1. Identify information needed and the service requested on a repair order.

2. Identify purpose and demonstrate proper use of fender covers, mats.

3. Demonstrate use of the three C's (concern, cause, and correction).

4. Review vehicle service history.

5. Complete work order to include customer information, vehicle identifying information, customer concern, related service history, cause, and correction.

Preparing Vehicle for Customer:

1. Ensure vehicle is prepared to return to customer per school/company policy (floor mats, steering wheel cover, etc.).

Objective

Upon completion of this job sheet, you will be able to prepare a service work order based on customer input, vehicle information, and service history. The student will also be able to describe the appropriate steps to take to protect the vehicle and delivering the vehicle to the customer after the repair.

Tools and Materials

An assigned vehicle or the vehicle of your choice

Service work order or computer-based shop management package

Parts and labor guide

Work Order Source: Describe the system used to complete the work order. If a paper repair order is being used, describe the source.

_____

_____

PROCEDURE

1. Prepare the shop management software for entering a new work order or obtain a blank paper work order. Describe the type of repair order you are going to use.

   _____

   _____

2. Enter customer information, including name, address, and phone numbers onto the work order.   Task Completed ☐

3. Locate and record the vehicle's VIN. Where did you find the VIN?

   _____

4. Enter the necessary vehicle information, including year, make, model, engine type and size, transmission type, license number, and odometer reading.   Task Completed ☐

5. Does the VIN verify that the information about the vehicle is correct?

   _____

6. Normally, you would interview the customer to identify his or her concerns. However, to complete this job sheet, assume the only concern is that the customer wishes to have the front brake pads replaced. Also, assume no additional work is required to do this. Add this service to the work order.   Task Completed ☐

7. Prepare the vehicle for entering the service department. Add floor mats, seat covers, and steering wheel covers to the vehicle.   Task Completed ☐

8. The history of service to the vehicle can often help diagnose problems, as well as indicate possible premature part failure. Gathering this information from the customer can provide some of the data needed. For this job sheet, assume the vehicle has not had a similar problem and was not recently involved in a collision. Service history is further obtained by searching files for previous service. Often this search is done by customer name, VIN, and license number. Check the files for any related service work.   Task Completed ☐

9. Search for technical service bulletins on this vehicle that may relate to the customer's concern. Did you find any? _____ If so, record the reference numbers here.

   _____

   _____

10. Based on the customer's concern, service history, TSBs, and your knowledge, what is the likely cause of this concern?

   _____

   _____

11. Add this information to the work order.   Task Completed ☐

12. Prepare to make a repair cost estimate for the customer. Identify all parts
    that may need to be replaced to correct the concern. List these here.

    _____

    _____

13. Describe the task(s) that will be necessary to replace the part.

    _____

    _____

14. Using the parts and labor guide, locate the cost of the parts that will be
    replaced and enter the cost of each item onto the work order at the appro-
    priate place for creating an estimate. If the valve or cam cover is leaking,
    what part will need to be replaced?

    _____

15. Now, locate the flat rate time for work required to correct the concern.
    List each task with its flat rate time.

    _____

    _____

16. Multiply the time for each task by the shop's hourly rate and add the cost
    of each item to the work order at the appropriate place for creating an
    estimate. Ask your instructor which shop labor rate to use and record
    it here.

    _____

17. Many shops have a standard amount they charge each customer for   Task Completed ☐
    shop supplies and waste disposal. For this job sheet, use an amount of
    ten dollars for shop supplies.

18. Add the total costs and insert the sum as the subtotal of the estimate.   Task Completed ☐

19. Taxes must be included in the estimate. What is the sales tax rate and does
    it apply to both parts and labor, or just one of these?

    _____

    _____

20. Enter the appropriate amount of taxes to the estimate, then add this to the Task Completed ☐
    subtotal. The end result is the estimate to give the customer.

21. By law, how accurate must your estimate be?

    _____

    _____

22. Generally speaking, the work order is complete and is ready for the customer's signature. However, some businesses require additional information; make sure you add that information to the work order. On the work order, there is a legal statement that defines what the customer is agreeing to. Briefly describe the contents of that statement.

_____

23. Now that the vehicle is completed and it is ready to be returned to the customer, what are the appropriate steps to take to deliver the vehicle to the customer? What would you do to make the delivery special? What should not happen when the vehicle is delivered to the customer?

_____

_____

Problems Encountered

_____

_____

_____

Instructor's Comments

_____

_____

_____

JOB SHEET

AUTOMATIC TRANSMISSIONS AND TRANSAXLES JOB SHEET 10

Road-Test a Vehicle to Check the Operation of an Automatic Transmission

Name _____ Station _____ Date _____

NATEF Correlation

This Job Sheet addresses the following **AST/MAST** task:

A.1. Identify and interpret transmission/transaxle concern, differentiate between engine performance and transmission/transaxle concerns; determine necessary action.

Objective

Upon completion of this job sheet, you will be able to identify and interpret transmission concerns, ensure proper engine operation, and diagnose transmission problems.

Tools and Materials

Service information

Clean rag

Paper and pencil

Scan tool

Describe the vehicle being worked on:

Year _____ Make _____ Model _____

VIN _____ Engine type and size _____

Model and type of transmission _____

PROCEDURE

Fluid Check with Dipstick

1. Park the vehicle on a level surface.   Task Completed ☐

2. Wipe all dirt off the protective disc and the dipstick handle.   Task Completed ☐

3. Start the engine and allow it to reach operating temperature.   Task Completed ☐

4. Remove the dipstick and wipe it clean with a lint-free cloth or paper towel.   Task Completed ☐

5. Reinsert the dipstick, remove it again, and record the fluid's level.

   _____

   _____

PROCEDURE

Fluid Check without Dipstick

1. Describe the process of checking the fluid level on your assigned vehicle.

   _____

   _____

2. Describe the condition of the fluid and tell what the fluid indicates.

   _____

   _____

3. Find and duplicate the chart from the service information that shows   Task Completed ☐
   the band and clutch application for the different gear selector positions.
   Using these charts will greatly simplify your diagnosis of a problem. It is
   also wise to have paper and a pencil to jot down notes about the operation
   of the transmission.

4. Inspect the transmission for leaks and other signs of wear or damage.
   What did you find?

   _____

   _____

5. Using a scan tool, check for any transmission DTCs that might be stored
   for the transmission, or any component that might affect transmission
   operation. Record your results here.

   _____

   _____

6. Drive the vehicle at normal speeds to warm up the engine and transmission. Describe the overall behavior of the transmission and torque converter.

   _____

   _____

7. Place the shift selector into the Drive or Overdrive position. Allow the transmission to shift through all of its gears. Describe the operation of the transmission and torque converter. If something sounded or felt abnormal during any of the shifts or while operating in a particular gear, refer to the band and clutch application chart to identify possible causes of the problem.

   _____

   _____

8. Force the transmission to downshift and record the quality of that downshift and the speed at which it downshifted.

   _____

   _____

9. Manually cause the transmission to downshift. Record the quality of that shift.

   _____

   _____

10. Stop the vehicle and put the shift selector in the lowest gear. Begin driving and manually shift the transmission through the forward gears. Record the quality of those shifts and your conclusions drawn from this drive. Again, if a problem was evident, refer to the band and clutch application chart to identify possible causes.

    _____

    _____

11. Stop the vehicle and place the gear selector into reverse. Describe the feel of that shift.

    _____

    _____

12. With the vehicle stopped, place the gear selector into park. Describe the feel of that shift.

    _____

    _____

13. What are your recommendations for this transmission?

    _____

    _____

Problems Encountered

_____

_____

_____

Instructor's Comments

_____

_____

_____

AUTOMATIC TRANSMISSIONS AND TRANSAXLES JOB SHEET 11

Gathering Vehicle Information

Name _____ Station _____ Date _____

NATEF Correlation

This Job Sheet addresses the following **MLR** task:

A.1. Research applicable vehicle and service information, fluid type, vehicle service history, service precautions, and technical service bulletins.

This Job Sheet addresses the following **AST/MAST** task:

A.2. Research applicable vehicle and service information, fluid type, vehicle service history, service precautions, and technical service bulletins.

Objective

Upon completion of this job sheet, you will be able to gather service information about a vehicle and its automatic transmission or transaxle.

Tools and Materials

Appropriate service information

Computer

Protective Clothing

Goggles or safety glasses with side shields

Describe the vehicle being worked on:

Year _____ Make _____ Model _____

VIN _____ Engine type and size _____

PROCEDURE

1. Using service information, describe what each letter and number in the VIN for this vehicle represents.

   _____

   _____

2. Locate the Vehicle Emissions Control Information (VECI) label and describe where you found it.

   _____

   _____

3. Summarize what information you found on the VECI label.

_____

_____

4. While looking in the engine compartment or under the vehicle, locate the identification tag on the transmission or transaxle. Describe where you found it.

_____

_____

5. Summarize the information contained on this label.

_____

_____

_____

6. Using the service information, locate the description of the vehicle's automatic transmission. List the major components of the control system and describe the basic operation parameters.

_____

_____

_____

_____

_____

7. Using the service information, locate and record all service precautions noted by the manufacturer regarding the automatic transmission.

_____

_____

_____

_____

8. Using the information available, locate and record the vehicle's service history.

_____

_____

_____

_____

9. Using the information sources available, summarize all Technical Service Bulletins for this vehicle that relate to the automatic transmission and its control system.

_____

_____

_____

_____

10. Refer to the service information and identify the exact type of fluid that should be used in this transmission. What type should be used?

_____

Problems Encountered

_____

_____

_____

Instructor's Comments

_____

_____

_____

AUTOMATIC TRANSMISSIONS AND TRANSAXLES JOB SHEET 12

Diagnosing Fluid Loss

Name _____ Station _____ Date _____

NATEF Correlation

This Job Sheet addresses the following **MLR** task:

A.4. Check transmission fluid condition; check for leaks.

This Job Sheet addresses the following **AST/MAST** task:

A.3. Diagnose fluid loss and condition concerns; determine necessary action.

Objective

Upon completion of this job sheet, you will be able to diagnose unusual fluid usage, level, and condition concerns.

Tools and Materials

Service information
Flashlight

Protective Clothing

Goggles or safety glasses with side shields

Describe the vehicle being worked on:

Year _____ Make _____ Model _____

VIN _____ Engine type and size _____

Model and type of transmission _____

PROCEDURE

1. Check the transmission housing for damage, cracks, and signs of leaks. Record your findings and recommendations.

   _____

   _____

2. Check the slip joint area in the transmission extension housing for leaks. If the transmission is a transaxle, check the area where the inner CV joint is attached to the housing. Record your findings and recommendations.

   _____

   _____

3. Check for leaks at all connections to the housing. Record your findings and recommendations.

_____

_____

4. Check the shift and throttle linkages for looseness, wear, and/or damage. Record your findings and recommendations.

_____

_____

5. Check the condition of all cooler lines and hoses. Record your findings and recommendations.

_____

_____

6. Check the condition of the fluid; pay attention to the level, color, smell, and feel of the fluid. Record your findings and recommendations.

_____

_____

Problems Encountered

_____

_____

_____

Instructor's Comments

_____

_____

_____

AUTOMATIC TRANSMISSIONS AND TRANSAXLES JOB SHEET 13

Checking the Fluid Level with a Dipstick

Name _____ Station _____ Date _____

NATEF Correlation

This Job Sheet addresses the following **MLR** task:

A.2. Check fluid level in a transmission or a transaxle equipped with a dip-stick.

This Job Sheet addresses the following **AST/MAST** task:

A.4. Check fluid level in a transmission or a transaxle equipped with a dip-stick.

Objective

Upon completion of this job sheet, you will be able to accurately determine the fluid level of an automatic transmission equipped with a dipstick.

Tools and Materials

Service information

Shop towel

Flashlight

Protective Clothing

Goggles or safety glasses with side shields

Describe the vehicle being worked on:

Year _____ Make _____ Model _____

VIN _____ Engine type and size _____

Model and type of transmission _____

PROCEDURE

1. Using service information, determine the proper transmission fluid for this transmission. List your results here.

   _____

   _____

2. What is the recommended fluid temperature when checking the transmission fluid level?

   _____

   _____

3. Is there a "cold" transmission fluid level checking procedure listed? If so, detail the procedure here.

_____

_____

_____

4. Describe the "hot" fluid level check next. If the manufacturer lists a temperature at which the fluid should be checked (using the scan tool) list it here as well.

_____

_____

_____

_____

_____

_____

_____

Problems Encountered

_____

_____

_____

Instructor's Comments

_____

_____

_____

AUTOMATIC TRANSMISSIONS AND TRANSAXLES JOB SHEET 14

Checking Fluid Level without a Dipstick

Name _____ Station _____ Date _____

NATEF Correlation

This Job Sheet addresses the following **MLR** task:

A.3. Check fluid level in a transmission or a transaxle not equipped with a dip-stick.

This Job Sheet addresses the following **AST/MAST** task:

A.5. Check fluid level in a transmission or a transaxle not equipped with a dip-stick.

Objective

Upon completion of this job sheet, you will be able to accurately determine the fluid level of an automatic transmission not equipped with a dipstick.

Tools and Materials

Service information

Shop towel

Flashlight

Protective Clothing

Goggles or safety glasses with side shields

Describe the vehicle being worked on:

Year _____ Make _____ Model _____

VIN _____ Engine type and size _____

Model and type of transmission _____

PROCEDURE

1. Using service information, determine the proper transmission fluid for this transmission. List your results here.

   _____

   _____

2. Is there a vent cap that has to be removed before checking the fluid level? Will the cap be replaced, or can it be reinstalled after use?

   _____

   _____

3. Describe the procedure to check the transmission fluid. Make sure to include the temperature at which the transmission fluid should be checked.

_____

_____

_____

_____

_____

_____

Problems Encountered

_____

_____

_____

Instructor's Comments

_____

_____

_____

AUTOMATIC TRANSMISSIONS AND TRANSAXLES JOB SHEET 15

Pressure Testing an Automatic Transmission

Name _____ Station _____ Date _____

NATEF Correlation

This Job Sheet addresses the following **MAST** task:

A.6. Perform pressure tests (including transmissions/transaxles equipped with electronic pressure control); determine necessary action.

Objective

Upon completion of this job sheet, you will have demonstrated the ability to conduct a pressure test on a transmission.

Tools and Materials

Hoist

Tachometer, as required   Lint-free shop towels

Pressure gauges with necessary adapters   Service information

Scan tool

Protective Clothing

Goggles or safety glasses with side shields

Describe the vehicle being worked on:

Year _____ Make _____ Model _____

VIN _____ Engine type and size _____

Model and type of transmission _____

PROCEDURE

1. Check the fluid level and condition before starting the test. Record your findings here.

   _____

2. List the pressure specifications with the required operating conditions.

   _____

   _____

3. Is the transmission fitted with electronic pressure control (EPC)?

   _____

4. Are there any DTCs that relate to transmission pressures stored? If so, record them here.

_____

_____

5. If the transmission has an EPC, what special procedures must be followed while checking the operating pressures of the transmission?

_____

_____

_____

_____

6. If the transmission is equipped with EPC, explain how the transmission pressure can be checked using a scan tool.

_____

_____

_____

_____

7. Have a fellow student sit in the vehicle to operate the throttle, brakes, and transmission during the test. Each vehicle tested may differ by make and model. Describe the procedure for your assigned vehicle.

_____

_____

_____

8. Raise the vehicle on the hoist to a comfortable working height.   Task Completed ☐

9. Describe the location of the pressure tap(s) on the transmission.

_____

_____

10. Connect the pressure gauges to the appropriate service ports.   Task Completed ☐

11. Start the engine. Have your helper press firmly on the brake pedal and apply the parking brake. The helper should then move the gear selector into the first test position.   Task Completed ☐

12. Run the engine at the specified speed. Move the gear selector as required by the service information.   Task Completed ☐

13. Observe and record the pressure readings at the various test conditions.

_____

_____

14. Turn off the engine and move the pressure gauges to the appropriate test ports for the next transmission range to be tested. Once again, some transmissions have only one pressure tap for line pressure, while some have more depending on make and model. Describe the location and purpose of the available pressure tap.

_____

_____

15. If your transmission has more than one pressure tap, move the gauge to the next tap position. Restart the engine and have your helper move the gear selector into the range to be tested, and then increase the engine's speed to the required test speed.

Task Completed ☐

16. Observe and record the pressure readings in the various test conditions (as applicable).

_____

_____

_____

17. Allow the engine to return to idle, and then turn it off.

Task Completed ☐

18. Repeat this sequence until all transmission ranges have been tested.

Task Completed ☐

19. Summarize the results of this test and compare them to the specifications.

_____

_____

_____

_____

_____

_____

20. Describe the importance of your findings as related to the diagnosis of this transmission.

_____

_____

_____

Problems Encountered

_____

_____

_____

Instructor's Comments

_____

_____

_____

AUTOMATIC TRANSMISSIONS AND TRANSAXLES JOB SHEET 16

Diagnosing Noises and Vibrations

Name _____ Station _____ Date _____

NATEF Correlation

This Job Sheet addresses the following **MAST** task:

A.7. Diagnose noise and vibration concerns; determine necessary action.

Objective

Upon completion of this job sheet, you will be able to diagnose transmission noise and vibration concerns.

Tools and Materials

A vehicle with an automatic transmission

Hoist

Stethoscope

Droplight or good flashlight

Service information

Protective Clothing

Goggles or safety glasses with side shields

Describe the vehicle being worked on:

Year _____ Make _____ Model _____

VIN _____ Engine type and size _____

PROCEDURE

1. Check the fluid level of the transmission before starting work. Record the results here.

   _____

   _____

2. Start the vehicle and leave in park. Is the vibration or noise present at idle?

   _____

   _____

3. Does the vibration or noise change with engine rpm? (if so go to number 4)

_____

4. To eliminate belt driven components, remove the drive belt(s) and run the engine at the same rpm that the noise is present. (Limit the time the engine is ran to a few seconds without the belt to prevent overheating). Did the noise or vibration go away?

_____

_____

5. If the vibration is still present with the belt removed, what could be the source of the problem?

_____

_____

6. Describe your findings here. Were you able to determine the source of the vibration or noise?

_____

_____

_____

Noise and/or Vibration while Driving

1. If the noise or vibration is present while driving, inspect the tires for any signs of wear that might produce a sound while driving, such as cupping, feathering, or broken belts. Look for obvious signs of a damaged rim. Describe your results here.

_____

_____

_____

2. Place the vehicle on the rack and report on the condition of the following:

   a. Engine and transmission mounts _____

   b. Broken transmission bell housing or missing bolts _____

   c. Cracked or broken flywheel _____

   d. Loose torque converter bolts _____

   e. Missing torque converter weights _____

   f. Damaged driveshaft or loose universal joints _____

Describe your findings.

_____

_____

_____

_____

3. If there is no obvious cause of a noise/vibration while inspecting the vehicle on the rack, carefully test drive the vehicle and listen for a change in the sound while driving over different types of pavement, such as asphalt and concrete. Is there a change in the sound? Record your results here.

_____

_____

_____

4. If the sound stays the same over different types of pavement, does the sound get louder with an increase in speed?

_____

5. To help determine if the sound is from a wheel bearing, see if the noise changes on turns. What did you find?

_____

6. If the noise appears to be from the driveline at a specific speed, try letting the vehicle coast down and then accelerate. Did the noise change with engine load?

_____

_____

7. If the noise/vibration does not change with engine load, observe the rpm at which the noise occurs. See if the noise occurs at the same rpm in different gears. For instance, if the noise occurs in fourth gear at 1800 rpm, does it occur at 1800 rpm in third gear?

_____

8. Did you locate the source of the noise or vibration? If so, give an overview of the problem here.

_____

_____

_____

Challenges Encountered

_____

_____

_____

Instructor's Comments

_____

_____

_____

AUTOMATIC TRANSMISSIONS AND TRANSAXLES JOB SHEET 17

Conduct a Stall Test

Name _____ Station _____ Date _____

NATEF Correlation

This Job Sheet addresses the following **AST** task:

A.6. Perform stall test; determine necessary action.

This Job Sheet addresses the following **MAST** task:

A.8. Perform stall test; determine necessary action.

Objective

Upon completion of this job sheet, you will be able to conduct a stall test.

Tools and Materials

A vehicle with an automatic transmission

Tachometer

Hoist

Stethoscope

Droplight or good flashlight

Service information

Protective Clothing

Goggles or safety glasses with side shields

Describe the vehicle being worked on:

Year _____ Make _____ Model _____

VIN _____ Engine type and size _____

PROCEDURE

> **WARNING:** *Some manufacturers do not recommend a stall test be conducted on their vehicles. Check the service information before proceeding. If a stall test should not be run on the assigned vehicle, conduct the test on a different vehicle. Make sure no one is standing in front of the vehicle during the stall test. Make certain the brakes are in good condition before performing the stall test. Do not allow the wheels to break traction during the stall test. Follow service information for limiting the duration of the stall test.*

1. Describe the transmission in the vehicle being tested.

_____

_____

2. Check and describe the condition and level of the fluid (color, condition, and smell).

   _____

   _____

3. What is indicated by the fluid's condition?

   _____

   _____

4. Connect a tachometer to the engine. If the vehicle has one in the instrument panel, there is no need to connect another one. Explain how you connected the tachometer.

   _____

   _____

5. Mark the face of the tachometer with a grease pencil at the recommended maximum rpm for this test. What is the maximum rpm? _____

6. Check the engine's coolant level. Describe your findings.

   _____

   _____

7. Block the front wheels and set the parking brake. Place the transmission in park and allow the engine and transmission to reach normal operating condition. How do you know when that temperature is reached?

   _____

   _____

8. Place the gear selector into the gear that is recommended for this test. What is the recommended gear? _____

9. Make sure no one is in front of the vehicle. Press the throttle pedal to the floor with your right foot and firmly press the brake pedal with your left. Hold the brake pedal down. When the tachometer needle stops rising, note the engine speed and let off on the throttle. What was the highest rpm during the test? (Complete the test as quickly as possible to prevent transmission damage.)

   _____

   _____

10. Place the gearshift into neutral and allow the engine to run at 1000 rpm for at least one minute. What is the purpose of doing this?

    _____

    _____

11. If noise is heard from the transmission during the stall test, raise the vehicle on a hoist. Describe the noise.

_____

_____

12. Use the stethoscope to determine if the noise is from the transmission or the torque converter. Describe your findings.

_____

_____

13. What are your conclusions from this test?

_____

_____

Problems Encountered

_____

_____

_____

Instructor's Comments

_____

_____

_____

AUTOMATIC TRANSMISSIONS AND TRANSAXLES JOB SHEET 18

Testing a Lock-up Converter

Name _____ Station _____ Date _____

NATEF Correlation

This Job Sheet addresses the following **AST** task:

A.7. Perform lock-up converter system tests; determine necessary action.

This Job Sheet addresses the following **MAST** task:

A.9. Perform lock-up converter system tests; determine necessary action.

Objective

Upon completion of this job sheet, you will be able to test the operation of a lock-up converter.

Tools and Materials

Service information

Scan tool

Digital multimeter

Protective Clothing

Goggles or safety glasses with side shields

Describe the vehicle being worked on:

Year _____ Make _____ Model _____

VIN _____ Engine type and size _____

PROCEDURE

1. Check with the service information to identify the lamp that warns the driver of a transmission concern. What lamp is it?

   _____

   _____

2. Start the engine and observe the above lamp and the engine MIL. What did you find?

   _____

   _____

3. Connect the scan tool and retrieve all DTCs. What were they?

_____

_____

4. If an engine-related problem is suspected, it should be corrected before moving on to the torque converter tests. Did any of the DTCs suggest an engine problem? If so, what was it?

_____

_____

5. Operate the vehicle with the scan tool connected. (Make sure to have a helper driving the vehicle, do not drive the vehicle and try to operate the scan tool.) Drive at 40 mph. With the scan tool, engage the lock-up converter's solenoid. Watch the tachometer while making the change. What did you observe? Engine speed should drop when the converter locks.

_____

_____

6. Disengage the clutch and observe the engine's speed. What did you observe?

_____

_____

7. The action of the lock-up converter can also be checked with the vehicle stopped and the engine running. Place the shifter into the "drive" position. With the scan tool, engage the lock-up solenoid. Describe what happened.

_____

_____

8. If there was no change in engine speed, what is evident?

_____

_____

9. The lock-up solenoid can also be checked with the ignition on but the engine off. Engage the solenoid with the scan tool and listen. Did you hear a click?

_____

_____

10. If a problem with the solenoid is suspected, remove it from the transmission for testing. How should it be tested?

_____

_____

11. Refer to the service manual and locate the resistance checks for the solenoid. What are the specifications?

_____

_____

12. Measure the resistance of the solenoid across the test points specified. What were your results?

_____

_____

13. What can you conclude from these tests?

_____

_____

Problems Encountered

_____

_____

_____

Instructor's Comments

_____

_____

_____

AUTOMATIC TRANSMISSIONS AND TRANSAXLES JOB SHEET 19

Logical Diagnosis

Name _____ Station _____ Date _____

NATEF Correlation

This Job Sheet addresses the following **AST** task:

A.8. Diagnose transmission/transaxle gear reduction/multiplication concerns using driving, driven, and held member (power flow) principles.

This Job Sheet addresses the following **MAST** task:

A.10. Diagnose transmission/transaxle gear reduction/multiplication concerns using driving, driven, and held member (power flow) principles.

Objective

Upon completion of this job sheet, you will be able to identify the most likely sources of a problem by using your understanding of the power flow of the transmission.

Tools and Materials

A vehicle with an automatic transmission

Service information

Protective Clothing

Goggles or safety glasses with side shields

Describe the vehicle being worked on:

Year _____ Make _____ Model _____

VIN _____ Engine type and size _____

PROCEDURE

NOTE: *Identifying what planetary gear set members are being used and which holding and apply devices are activated in each gear is not merely an exercise to keep you busy. Rather, the understanding of power flow through a transmission is an extremely valuable tool for diagnostics. This job sheet is designed to make you think about the power flow and the transmission parts involved while the transmission is operating in a particular gear range. This job sheet is not completed on a vehicle, rather it should be completed with you sitting down and thinking and applying your understanding of planetary gear action and the power flow of a particular transmission to answer these questions.*

1. Describe the transmission in the vehicle being tested (include the number of forward gears, clutches, bands, and electronic controls).

   _____

   _____

   _____

   _____

2. When the engine is running and the transmission is in Park, what parts of the transmission are rotating with the engine?

   _____

   _____

3. When the transmission is in Park, what parts could be the cause of a noise or vibration?

   _____

   _____

4. If Park does not properly engage and hold, what are the possible problems?

   _____

   _____

5. When the engine is running and the transmission is in Neutral, what parts of the transmission are rotating with the engine?

   _____

   _____

6. When the transmission is in Neutral, what parts could be the cause of a noise or vibration?

   _____

   _____

7. When the engine is running and the transmission is in Drive, what parts of the transmission are rotating with the engine during all forward gear ranges?

   _____

   _____

8. When the transmission is in Drive, what parts could be the cause of a noise or vibration?

   _____

   _____

9. When the engine is running and the transmission is in First gear, what parts of the transmission are rotating with the engine?

_____

_____

10. When the transmission is in First gear, what parts could be the cause of a noise or vibration?

_____

_____

11. When the transmission is in First gear, what could be the cause of engagement problems?

_____

_____

12. When the engine is running and the transmission is in Second gear, what parts of the transmission are rotating with the engine?

_____

_____

13. When the transmission is in Second gear, what parts could be the cause of a noise or vibration?

_____

_____

14. When the transmission is in Second gear, what could be the cause of engagement problems?

_____

_____

15. When the engine is running and the transmission is in Third gear, what parts of the transmission are rotating with the engine?

_____

_____

16. When the transmission is in Third gear, what parts could be the cause of a noise or vibration?

_____

_____

17. When the transmission is in Third gear, what could be the cause of engagement problems?

_____

_____

18. When the engine is running and the transmission is in Fourth gear, what parts of the transmission are rotating with the engine?

_____

_____

19. When the transmission is in Fourth gear, what parts could be the cause of a noise or vibration?

_____

_____

20. When the transmission is in Fourth gear, what could be the cause of engagement problems?

_____

_____

21. When the engine is running and the transmission is in Fifth gear, what parts of the transmission are rotating with the engine?

_____

_____

22. When the transmission is in Fifth gear, what parts could be the cause of a noise or vibration?

_____

_____

23. When the transmission is in Fifth gear, what could be the cause of engagement problems?

_____

_____

24. When the engine is running and the transmission is in Reverse gear, what parts of the transmission are rotating with the engine?

_____

_____

25. When the transmission is in Reverse gear, what parts could be the cause of a noise or vibration?

_____

_____

26. When the transmission is in Reverse gear, what could be the cause of engagement problems?

_____

_____

Problems Encountered

_____

_____

_____

Instructor's Comments

_____

_____

_____

Problems Encountered

Instructor's Comments

AUTOMATIC TRANSMISSIONS AND TRANSAXLES JOB SHEET 20

Diagnose Control Systems

Name _____ Station _____ Date _____

NATEF Correlation

This Job Sheet addresses the following **MAST** task:

A.11. Diagnose electronic transmission/transaxle control systems using appropriate test equipment and service information.

Objective

Upon completion of this job sheet, you will be able to diagnose electronic, mechanical, and hydraulic control system concerns.

Tools and Materials

Component locator   Digital multimeter

Small mirror   Flashlight

Scan tool

Protective Clothing

Goggles or safety glasses with side shields

Describe the vehicle being worked on:

Year _____ Make _____ Model _____

VIN _____ Engine type and size _____

PROCEDURE

1. Check with the service information to identify the lamp that warns the driver of a transmission concern. What lamp is it?

   _____

   _____

2. Start the engine and observe the previous step's lamp and the engine MIL. What did you observe?

   _____

   _____

3. Connect the scan tool and retrieve all DTCs. What were they?

   _____

   _____

4. If an engine-related problem is suspected, it should be corrected before moving on to the torque converter tests. Did any of the DTCs suggest an engine problem? If so, what.

_____

_____

5. Check all electrical connections to the transmission. On transaxles, the connectors can normally be inspected through the engine compartment, whereas they can only be seen from under the vehicle on longitudinally mounted transmissions. Make sure they are tight and not damaged. Record your findings.

_____

_____

6. Release the locking tabs of the connectors and disconnect them, one at a time, from the transmission. Carefully examine them for signs of corrosion, distortion, moisture, and transmission fluid. A connector or wiring harness may deteriorate if ATF reaches it. Also, check the connector at the transmission. Using a small mirror and flashlight may help you get a good look at the inside of the connectors. Record your findings.

_____

_____

7. Inspect the entire transmission wiring harness for tears and other damage. Record your findings.

8. Because the operation of the engine and transmission are integrated through the control computer, a faulty engine sensor or connector may affect the operation of both. The engine control sensors that are the most likely to cause shifting problems are the throttle-position sensor, MAP sensor, and vehicle speed sensor. Locate these sensors and describe their location.

_____

_____

9. Remove the electrical connector from the TP sensor and inspect both ends for signs of corrosion and damage. Poor contact can cause the transmission to miss shifts. Record your findings.

_____

_____

10. Inspect the wiring harness to the TP sensor for evidence of damage. Record your findings.

     _____

     _____

11. Check both ends of the three-pronged connector and wiring at the MAP sensor for corrosion and damage. Record your findings.

     _____

     _____

12. Check the condition of all vacuum hoses. Record your findings.

     _____

     _____

13. Check the speed sensor's connections and wiring for signs of damage and corrosion. Record your findings.

     _____

     _____

14. Record your conclusions from the visual inspection.

     _____

     _____

15. Check the service information and identify the sensors located on the transmission. List them here along with their function.

     _____

     _____

     _____

     _____

16. Most late-model transmissions use shift solenoids to control shifting. What is the recommended method of testing the shift solenoids?

     _____

     _____

     _____

     _____

     _____

17. Using service information, explain the process of shifting the transmission by using a special tool or scan tool.

_____

_____

_____

_____

_____

_____

18. Electronically controlled transmissions rely on electrical signals from a speed sensor to control shift timing. The speed sensor used in many late-model transmissions is a permanent magnetic (PM) generator. Locate the speed sensor and describe its location.

_____

_____

19. Raise the vehicle on a lift. Allow the wheels to be suspended and rotate freely. Set your DMM to measure AC voltage. Connect the meter to the speed sensor. Start the engine and put the transmission in gear. Slowly increase the engine's speed until the vehicle is at approximately 20 mph, and then measure the voltage at the speed sensor. Record your findings.

_____

_____

20. Slowly increase the engine's speed and observe the voltmeter. The voltage should increase smoothly and precisely with an increase in speed. Record your findings.

_____

_____

21. A speed sensor can also be tested with it out of the vehicle. Connect an ohmmeter across the sensor's terminals. What was the measured resistance?

_____

_____

22. Locate the specifications for the sensor and compare your readings with specifications.

_____

_____

23. Determine if your transmission is equipped with an input speed sensor. If it is, explain the purpose of the input speed sensor and the important function it has in transmission control and monitoring.

_____

_____

Problems Encountered

_____

_____

_____

Instructor's Comments

_____

_____

_____

AUTOMATIC TRANSMISSIONS AND TRANSAXLES JOB SHEET 21

Using Pascal's Law to Help with Diagnostics

Name _____ Station _____ Date _____

NATEF Correlation

This Job Sheet addresses the following **AST** task:

A.9. Diagnose pressure concerns in a transmission using hydraulic principles (Pascal's Law).

This Job Sheet addresses the following **MAST** task:

A.12. Diagnose pressure concerns in the transmission using hydraulic principles (Pascal's law).

Objective

Upon completion of this job sheet, you will be able to apply the basic principles of Pascal's law to operation of an automatic transmission and to the results of a pressure test.

Tools and Materials

A good text book

Describe the vehicle being worked on:

Year _____ Make _____ Model _____

VIN _____ Engine type and size _____

PROCEDURE

1. The hydraulic operation of an automatic transmission is based on the principles of Pascal's law. Briefly describe how this law applies to the operation of automatic transmissions.

   _____

   _____

2. Briefly explain why incorrect fluid pressures caused by a bad oil pump would affect all operating modes of a transmission.

   _____

   _____

3. The transmission is fitted with several pressure-regulating and flow-directing valves. Briefly describe what they do to the movement of fluid throughout the transmission.

   _____

   _____

4. A servo assembly is used to control the application of a band, which must tightly hold the drum it surrounds when it is applied. The holding capacity of the band is determined by the construction of the band and the pressure applied to it. What does the servo do?

_____

_____

5. If a servo has an area of 10 square inches and has a pressure of 70 psi applied to it, the apply force of the servo will be _____ pounds.

6. A multiple-disc assembly is also used to stop and hold gear set members. If the fluid pressure applied to the clutch assembly is 70 psi and the diameter of the clutch piston is 5 inches, the force applying the clutch pack is _____ pounds.

7. List three probable causes for all fluid pressures to be low during a pressure test and explain why.

_____

_____

_____

8. What could possibly be indicated by a great pressure drop between shifts?

_____

_____

9. Why are transmissions checked at various engine speeds and throttle openings?

_____

_____

10. If there is an internal leak in the Second gear circuit of the transmission, how would this affect its operation?

_____

_____

11. If the line pressure is too high, how will the operation of the transmission be affected?

_____

_____

Problems Encountered

_____

_____

_____

Instructor's Comments

_____

_____

_____

AUTOMATIC TRANSMISSIONS AND TRANSAXLES JOB SHEET 22

Servicing Linkages

Name _____ Station _____ Date _____

NATEF Correlation

This Job Sheet addresses the following **MLR/AST/MAST** task:

B.1. Inspect, adjust, and replace manual valve shift linkage, transmission range sensor/switch, and park/neutral position switch.

Objective

Upon completion of this job sheet, you will be able to inspect, adjust, or replace throttle valve and gear selector linkages and cables.

Tools and Materials

Basic hand tools

Service information

Protective Clothing

Goggles or safety glasses with side shields

Describe the vehicle being worked on:

Year _____ Make _____ Model _____

VIN _____ Engine type and size _____

PROCEDURE

CAUTION: *Always set the parking brake before moving the gear selector through its positions.*

1. If the gear selector linkage is misadjusted, poor gear engagement, slipping, and excessive wear can result. The gear selector linkage should be adjusted so the manual shift valve detent position in the transmission matches the selector level detent and position indicator. To check the adjustment of the linkage, move the shift lever from the Park position to the lowest Drive gear. Detents should be felt at each of these positions. If the detent cannot be felt in either of these positions, the linkage needs to be adjusted. Describe your findings.

_____

_____

2. Move the gear selector lever slowly until it clicks into the Park position. Turn the ignition key to the Start position. If the starter operates, the Park position is correct. After checking the Park position, move the lever slowly toward the Neutral position until the lever drops at the end of the N stop in the selector gate. If the starter also operates at this point, neutral is okay. By completing this routine, what did you check?

_____

_____

3. If the engine does not start in P or N, the transmission range switch might be incorrectly adjusted or defective. The signal from the transmission range switch can be checked with the scan tool for validity, as well as trouble codes. Describe the scan tool data for the range switch and the functions the switch performs.

_____

_____

_____

_____

4. Outline the procedure to adjust the transmission range switch.

_____

_____

_____

_____

5. If your vehicle is equipped with a floor-mounted gear selector, outline the steps to adjust the gear selector linkage.

_____

_____

_____

_____

_____

6. If your vehicle is equipped with a cable-mounted linkage, outline the steps to adjust the gear selector linkage.

_____

_____

_____

Did you have any problems?

_____

_____

7. On some vehicles, you may need to adjust the neutral safety switch after resetting the linkage. Did you need to do this? Why?

_____

_____

8. If you are unable to make an adjustment, the levers' grommets may be badly worn or damaged and should be replaced. When it is necessary to disassemble the linkage from the levers, the plastic grommets used to retain the cable or rod should be replaced. Use a prying tool to force the cable or rod from the grommet, and then cut out the old grommet. Pliers can be used to snap the new grommets into the levers and the cable or rod into the levers. What did you find?

_____

Problems Encountered

_____

_____

_____

Instructor's Comments

_____

_____

_____

AUTOMATIC TRANSMISSIONS AND TRANSAXLES JOB SHEET 23

Inspecting Gaskets, Seals, and Bushings

Name _____ Station _____ Date _____

NATEF Correlation

This Job Sheet addresses the following **MLR/AST/MAST** task:

B.2. Inspect for leakage at external seals, gaskets, and bushings.

Objective

Upon completion of this job sheet, you will be able to inspect common seals, gaskets, and bushings in a transmission/transaxle.

Tools and Materials

Service information

Seal removal and installation tools

Clean ATF

Hand file

Puller and driver set

Protective Clothing

Goggles or safety glasses with side shields

Describe the vehicle being worked on:

Year _____ Make _____ Model _____

VIN _____ Engine type and size _____

PROCEDURE

Inspection

1. With the vehicle on a lift, check for leaks at and around the transmission's extension housing. Record your findings.

   _____

   _____

2. An oil leak stemming from the mating surfaces of the extension housing and the transmission case may be caused by loose bolts. To correct this problem, tighten the bolts to the specified torque. What is the specified torque?

   _____

3. Check the extension housing for cracks, especially around the case mounting surface and the pad that attaches to the transmission mount. Record your findings.

_____

_____

4. Also check for signs of leakage at the rear of the extension housing. Fluid leaks from the seal of the extension housing can be corrected with the transmission in the car. Record your findings.

_____

_____

Seal and Bushing Replacement

1. If the leak was at the rear of the housing, the problem may be the bushing or seal. The clearance between the drive shaft's sliding yoke and the bushing should be minimal. If the clearance is satisfactory, a new oil seal will correct the leak. If the clearance is excessive, a new seal and a new bushing should be installed.

_____

_____

2. To begin the procedure for replacing the bushing and seal, remove the drive shaft from the vehicle. What do you need to remember before removing the drive shaft?

_____

_____

3. Insert the appropriate puller tool into the extension housing until it grips the front side of the bushing.   Task Completed ☐

4. Pull the seal and bushing from the housing.   Task Completed ☐

5. To replace the bushing, drive a new bushing, with the appropriate driver, into the extension housing. Always make sure this bushing is aligned correctly during replacement, otherwise premature failure can result.   Task Completed ☐

6. Lubricate the lip of the seal, and then install the new seal in the extension housing. What did you use to install the seal?

_____

7. Install the drive shaft.   Task Completed ☐

Extension Housing Seal

1. If only the seal needs to be replaced, remove the drive shaft and then remove the old seal with a puller.

Task Completed ☐

2. Lubricate the lip of the seal, and then install the new seal in the extension housing. What did you use to install the seal?

_____

3. Install the drive shaft.

Task Completed ☐

Vehicle Speed Sensor VSS Seal

1. An oil leak at the VSS can be corrected by replacing the O-ring seal. Check the seal and record the results.

_____

_____

2. If the seal is bad, remove the VSS assembly from the extension housing by removing the hold-down screw that keeps the retainer in its bore.

Task Completed ☐

3. Carefully pull up on the VSS to remove it from the transmission.

Task Completed ☐

4. Carefully remove the old seal from the VSS.

Task Completed ☐

5. Lightly grease and install the O-ring onto the retainer.

Task Completed ☐

6. Install the hold-down screw and tighten it in place.

Task Completed ☐

Extension Housing Service

1. If the cause of the leak is not due to a faulty seal or bushing, prepare to remove the extension housing by removing the drive shaft.

Task Completed ☐

2. Unbolt the extension housing from the transmission.

Task Completed ☐

3. Clean the extension housing.

Task Completed ☐

4. Using a known flat surface, check the flatness of the mating surface. Any defects that cannot be removed by light filing indicate that the housing should be replaced. Record your findings.

_____

_____

5. Carefully inspect all bores, whether they are threaded or not. All damaged threaded areas should be repaired. If any condition exists that cannot be adequately repaired, the housing should be replaced. Record your findings.

_____

_____

6. Before installing the rear extension housing, assemble the parking pawl pin, washer, spring, and pawl, and any other assembly that is enclosed by the extension housing. Are they assembled properly and in good shape?

   _____

   _____

7. Install a new extension housing gasket and tighten the housing to the transmission. What are the torque specifications for those bolts?

   _____

   _____

8. Reinstall the drive shaft.   Task Completed ☐

General Seals

1. Identify the main components of an automatic transmission that have seals that should be replaced during an overhaul and list them here.

   _____

   _____

   _____

   _____

2. Assuming these will be replaced, make sure the replacement seals are of the type that is recommended by the manufacturer of the transmission.   Task Completed ☐

3. Check each seal before installing them in their own bore. They should be slightly smaller or larger (3%) than their groove or bore. Do not assume that because a particular seal came with the overhaul kit it is the correct one. What did you find?

   _____

   _____

4. After lubricating them, keep the seals clean and free of dirt. Before installing seals, clean the shaft and/or bore area.   Task Completed ☐

5. Carefully check the shaft and/or bore area for damage. File or stone away any burrs or bad nicks and polish the surfaces with a fine crocus cloth. Then, clean the area to remove the metal particles.   Task Completed ☐

6. Lubricate all seals in clean ATF, unless otherwise instructed by the manufacturer. Remember to use only the proper fluids as stated in the appropriate service manual.   Task Completed ☐

7. O-ring and square-cut seals are used to seal non-rotating parts. These seals are commonly used on oil pumps, servos, clutch pistons, speedometer drives, and vacuum modulators. Where will you install this type of seal?

   _____

   _____

8. Coat the entire seal with assembly lube.   Task Completed ☐

9. Install the seal while making sure you do not stretch or distort the seal while you work it into its holding groove. After a square-cut seal is installed, double-check it to make sure it is not twisted. The flat surface of the seal should be parallel with the bore. Did you have any problems?

   _____

   _____

10. Lip seals are used around rotating shafts and apply pistons. The rigid outer diameter provides a mounting point for the seal and is pressed into a bore. What seal will you be replacing?

   _____

   _____

11. Set the seal in place. Make sure the lip is facing the correct direction. Use   Task Completed ☐
    the correct driver to install the seal and be careful not to damage the seal during installation.

12. Teflon or metal sealing rings are commonly used to seal servo pistons, oil pump covers, and shafts. They are designed to provide a seal; they may also allow a controlled amount of fluid leakage to lubricate a bushing. What type of seal will you be installing for this job sheet?

   _____

   _____

13. Solid sealing rings are commonly made of a Teflon-based material and never reused. To remove them, cut the seal after it has been pried out of its groove. Installing a new ring requires special tools. What tools do you have to do this?

   _____

   _____

14. What will happen if you install this type of seal without the proper tools?

   _____

   _____

15. Open-end sealing rings fit loosely into a machined groove. This type of ring is typically removed and installed with a pair of snap ring pliers. Do you have the appropriate pliers?

   _____

   _____

16. Butt-end sealing rings can be removed with a small screwdriver. To install this type of ring, use a pair of snap ring pliers and expand the ring to move it into position. Do you have the appropriate pliers?

   _____

   _____

17. Locking-end rings may have hooked ends that connect or ends that are cut at an angle. These seals are removed and installed in the same way as butt-end rings. After these rings are installed, make sure the ends are properly positioned and touching. Where do these seals fit into the transmission you are working on?

_____

_____

18. Check all metal sealing rings for proper fit. Insert them in their bore. Do they feel tight?

_____

_____

19. Check the fit of the rings in their shaft groove. If the ring can move laterally in the groove, the groove is worn. What did you find?

_____

_____

20. Check the side clearance of the ring by placing the ring into its groove and measure the clearance between the ring and the groove with a feeler gauge. Typically, the clearance should not exceed 0.003 inches. What did you find?

_____

_____

21. Before installing the seals, look for nicks in the grooves and for evidence of groove taper or stepping. If the grooves are tapered or stepped, the shaft will need to be replaced. If there are burrs or nicks in the grooves, they can be filed away. What did you find?

_____

_____

22. Install the seals. If the seal has some form of locking ends, these should be interlocked prior to trying the seal in its bore.

Task Completed ☐

General Gaskets

1. Gaskets are used to seal non-moving components. However, their sealing ability is critical to the durability of a transmission. How many gaskets do you have in your overhaul kit?

_____

_____

2. Make sure all gasket surfaces are clean and flat. Transmission gaskets should not be installed with any type of liquid adhesive or sealant, unless specifically noted by the manufacturer. If any sealer gets into the valve body, severe damage can result. Also, sealant can clog the oil filter. If a gasket is difficult to install, a thin coating of transmission assembly lube can be used to hold the gasket in place.

Task Completed ☐

3. One of the common locations for a gasket is between the main oil pump and the transmission case. Will you be installing a new gasket here?

_____

_____

4. If so, place some petroleum jelly in two or three spots around the oil pump gasket and position it on the housing.

Task Completed ☐

5. Align the pump and install it with care. Tighten the pump attaching bolts to specifications in the specified order. What are those specifications?

_____

_____

6. Another common location for a gasket is the oil pan. Oil pans are typically made of stamped steel and can become distorted around the attaching bolt holes. Carefully inspect the sealing surface and describe what you found and what needs to be done to the pan.

_____

_____

Problems Encountered

_____

_____

_____

Instructor's Comments

_____

_____

_____

AUTOMATIC TRANSMISSIONS AND TRANSAXLES JOB SHEET 24

Testing Electrical/Electronic Components

Name _____ Station _____ Date _____

NATEF Correlation

This Job Sheet addresses the following **AST/MAST** task:

B.3. Inspect, test, adjust, repair, or replace electrical/electronic components and circuits, including computers, solenoids, sensors, relays, terminals, connectors, switches, and harnesses.

Objective

Upon completion of this job sheet, you will be able to inspect and test, adjust, repair, or replace transmission-related electrical and electronic components.

Tools and Materials

Digital multimeter   Test light

Service information   Self-powered test light

Scan tool

Protective Clothing

Goggles or safety glasses with side shields

Describe the vehicle being worked on:

Year _____ Make _____ Model _____

VIN _____ Engine type and size _____

Transmission type and model _____

PROCEDURE

1. Using service information and the component locator for the vehicle, list all of the electrical and electronic components located on the transmission.

   _____

   _____

   _____

2. Carefully check all electrical wires and connectors for damage, looseness, and corrosion. Record your findings.

   _____

   _____

3. Check all ground cables and connections. Corroded battery terminals and/or broken or loose ground straps to the frame or engine block will cause problems. Record your findings.

_____

_____

4. Check the fuse or fuses to the control module. To accurately check a fuse, either test it for continuity with an ohmmeter or check each side of the fuse for power when the circuit is activated. Record your findings.

_____

_____

5. Make sure the battery's and the alternator's output voltage is at least 12.6 volts. If a problem is found here, correct it before continuing to diagnose the system. Record your findings.

_____

_____

6. With the system on, measure the voltage dropped across connectors and circuits. Record your findings.

_____

_____

7. Prepare the vehicle for a road test. Have a notebook handy to record events during the test. Connect the scan tool to the DLC. Have an assistant drive the vehicle while you operate the scan tool. Do not drive the vehicle and operate the scan tool at the same time. All pressure and gear changes should be noted. Also, the various computer inputs should be monitored and the readings recorded for future reference. Use the snapshot function of the scan tool to record the data present when the transmission appears to be malfunctioning. Summarize the results of the road test.

_____

_____

8. After completing the road test, use the scan tool to retrieve trouble codes and review the snapshot data recorded. Explain your findings.

_____

_____

_____

9. Referring to the service information, describe what is indicated by the trouble codes (if applicable).

_____

_____

10. If a problem with shift points was noted during the road test, make sure the transmission is not operating in its default mode. Record your findings.

_____

_____

11. If the self-test sequence pointed to a problem in an input circuit, the input should be tested to determine the exact malfunction. Briefly describe the procedure outlined by the manufacturer.

_____

_____

12. Switches can be tested for operation and for excessive resistance with a voltmeter, test light, or ohmmeter. To check the operation of a switch with a voltmeter or a test light, connect the meter's positive lead to the battery side of the switch. With the negative lead attached to a good ground, the voltage should be measured at this point. Is this test possible? If not, why?

_____

_____

13. Without closing the switch, move the positive lead to the other side of the switch. If the switch is open, no voltage will be present at that point. The amount of voltage present at this side of the switch should equal the amount on the other side when the switch is closed. If the voltage decreases, the switch is causing a voltage drop due to excessive resistance. If no voltage is present on the ground-side of the switch with it closed, the switch is not functioning properly and should be replaced. Record your findings.

_____

_____

14. If a switch has been removed from the circuit, it can be tested with an ohmmeter or a self-powered test light. By connecting the leads across the switch connections, the action of the switch should open and close the circuit. Record your findings.

_____

_____

15. The manual lever switch is open or closed, depending on the position, and can be checked with an ohmmeter. By referring to a wiring diagram, you should be able to determine when the switch should be open. Then, connect the meter across the input and output of the switch. Move the lever into the desired position and measure the resistance. An infinite reading is expected when the switch is open. If there is any resistance, the switch should be replaced. Record your findings.

_____

_____

16. The pressure switches used in today's transmissions are either grounding switches or they connect or disconnect two wires. Refer to the wiring diagram to determine the type of switch and test the switch with an ohmmeter. Base your expected results on the type of switch you are testing. By using air pressure, you can easily see if the switch works properly or if it has a leak. Record your findings.

   _____

   _____

17. A type of sensor commonly used is a potentiometer. Rather than open and close a circuit, a potentiometer controls the circuit by varying its resistance in response to something. A TP sensor is a potentiometer that sends very low voltage back to the computer when the throttle plates are closed and increases the voltage as the throttle is opened. Most TP sensors receive a reference voltage of 5 volts. What the TP sensor sends back to the computer is determined by the position of the throttle. A TP sensor on most late-model vehicles is best checked by using a scan tool or oscilloscope. You should be able to watch the resistance across the TP sensor change as the throttle is opened and closed. Compare your reading to the specification. Record your findings.

   _____

   _____

   _____

18. With a voltmeter, you will be able to measure the reference voltage and the output voltage. Both of these should be within specified amounts. If the reference voltage is lower than normal, check the voltage drop across the reference voltage circuit from the computer to the TP sensor. Record your findings.

   _____

   _____

19. Vehicle speed sensors provide road speed information to the computer. Most VSS sensors are AC voltage generators and rely on a stationary magnet wrapped with a coil of wire. A rotating shaft fitted with iron teeth is located close to the sensor. Each time a tooth passes through the magnetic field, an electrical pulse is present. By counting the number of teeth on the output shaft, you can determine how many pulses per revolution you will measure with a voltmeter set to AC volts. How many teeth are on the output shaft? How many pulses on the meter did you observe?

   _____

   _____

   _____

20. A mass-airflow sensor is used to determine engine load by measuring the mass of the air being taken into the throttle body. The mass airflow sensor is a wire located in the intake air stream that receives a fixed voltage. The wire is designed so that it changes resistance in response to temperature changes. This sensor can be measured with a Digital multimeter set to the Hz frequency range. Check the service information for specific values. Summarize the specifications for this and record what you observed while checking this sensor.

_____

_____

_____

21. A scan tool can also be used to test a mass-airflow sensor. While diagnosing these systems, keep in mind that cold air is denser than warm air. Record your findings.

_____

_____

22. Temperature sensors are designed to change resistance with changes in temperature. A temperature sensor is based on a thermistor. Some thermistors increase resistance with an increase of temperature. Others decrease the resistance as temperature increases. Obviously, these sensors can be checked with an ohmmeter. To do so, remove the sensor. Then, determine the temperature of the sensor and measure the resistance across it. Compare your reading with the chart of normal resistances given in the service manual. Was the sensor you tested a PTC or an NTC?

_____

_____

_____

23. The controls of an electronically controlled transmission direct hydraulic flow through the use of solenoid valves. If you were unable to identify the cause of a transmission problem through the previous checks, you should continue your diagnostics by testing the solenoids. Before continuing, you must first determine if the solenoids are case grounded and fed voltage by the computer, or if they always have power applied to them and the computer merely supplies the ground. While looking in the service information to find this, also find the area that tells you which solenoids are on and which are off for each of the different gears. Summarize how the solenoids in your transmission work and how they are switched for the various speed gears and other functions.

_____

_____

_____

_____

24. To begin this test, you should connect a scan tool to the DLC that has the capability to control the transmission shifting.

_____

_____

_____

25. Increase your speed, and then turn on the solenoid for second gear. The transmission should have immediately shifted into second gear. Record your findings.

_____

_____

26. When you want to shift into third, select the combination of solenoid off and on to provide that gear. Record your findings.

_____

_____

27. Do the same for fourth and fifth gears. Record your findings.

_____

_____

28. Solenoids can be checked for circuit resistance and shorts to ground. The test can be conducted at the transmission case connector. By identifying the proper pins in the connector, individual solenoids can be checked with an ohmmeter. Record your findings.

_____

_____

29. Solenoids can also be tested on a bench. Resistance values are typically given in the service information for each application. Summarize the resistance specifications and the results of your test.

_____

_____

30. A solenoid may be electrically fine but still may fail mechanically or hydraulically. A solenoid's check valve may fail to seat or the porting can be plugged. To check for this, listen to the solenoid when it is activated. When a solenoid affected in this way is activated, it will make a slow dull thud. A good solenoid tends to snap when activated. Record your findings.

_____

_____

31. Describe how line pressure is controlled by your assigned transmission.

_____

_____

32. Describe a cause of high line pressure.

_____

33. Describe a cause of low line pressure.

_____

34. Using the scan tool, can you command line pressures? _____

How could this help you in your diagnostics?

_____

_____

35. Using the scan tool, see if any "shift adapts" are being used for this transmission and record them here:

_____

36. What do shift adapts indicate?

_____

Problems Encountered

_____

_____

_____

Instructor's Comments

_____

_____

_____

AUTOMATIC TRANSMISSIONS AND TRANSAXLES JOB SHEET 25

Visual Inspection and Filter and Fluid Change

Name _____ Station _____ Date _____

NATEF Correlation

This Job Sheet addresses the following **MLR/AST/MAST** task:

B.4. Drain and replace fluid and filter(s).

Objective

Upon completion of this job sheet, you will be able to conduct a preliminary inspection of an automatic transmission or transaxle and replace its fluid and filter.

Tools and Materials

Lift   Lint-free rags

Large drain pan   Pound-inch torque wrench

Protective Clothing

Goggles or safety glasses with side shields

Describe the vehicle being worked on:

Year _____ Make _____ Model _____

VIN _____ Engine type and size _____

Transmission type and model _____

PROCEDURE

1. Properly position the vehicle on a lift. Check the transmission housing for damage, cracks, and signs of leaks. Record your findings.

   _____

   _____

2. Check the end of the extension housing for signs of leakage. Record your findings.

   _____

   _____

3. Check all wiring and mechanical linkages for looseness or damage. Record your findings.

   _____

   _____

4. Check the condition of the cooler lines and hoses. Record your findings.

   _____

   _____

5. Check the condition of the fluid. Check its level, color, smell, and feel. Record your findings.

   _____

   _____

6. Remove or move any part of the vehicle that may interfere with the removal of the oil pan. What did you need to move?

   _____

   _____

   _____

7. Position the oil drain pan under the transmission pan. Loosen all the pan bolts and remove all but three at one end. Why should you do this?

   _____

   _____

8. After most of the fluid has drained from the pan, support the pan with one hand. Now, remove the remaining bolts and pour the rest of the transmission fluid into the drain pan. Carefully inspect the oil pan and the residue in it. Record your findings. What is indicated by what was in the pan?

   _____

   _____

   _____

9. Remove the old pan gasket and wipe the pan clean with a clean lint-free rag. Unbolt the fluid filter from the transmission's valve body. Keep the drain pan under the transmission while doing this. Did you have any difficulties? What were they?

   _____

   _____

   _____

10. Gather the new filter and gaskets. Compare the old with the new. What did you find? If there is a difference, what should you do?

    _____

    _____

11. Install the new filter onto the valve body and tighten the attaching bolts to the proper specifications. What are the specifications?

    _____

12. Lay the new pan gasket over the sealing area of the oil pan. Make sure the holes line up properly. Did they?

    _____

13. Position the pan onto the transmission. Install the attaching bolts and hand-tighten each. Tighten the bolts to the specified torque. Make sure you stagger the tightening. What are the specifications? What order did you follow when tightening the bolts?

    _____

    _____

14. Lower the vehicle. Pour a little less than the required amount of the recommended fluid into the transmission. How much fluid did you put in and what type of fluid was it?

    _____

15. Start the engine. Look under the vehicle and check for leaks. Was there any sign of leaks?

    _____

    _____

16. With the parking brake applied and the brake pedal depressed, move the gear selector through the gears. After the engine reaches normal operating temperature, put the transmission into park. Check the fluid level and correct it if necessary. What did you need to do to check the fluid level on this transmission?

    _____

    _____

    _____

Problems Encountered

_____

_____

_____

Instructor's Comments

_____

_____

_____

AUTOMATIC TRANSMISSIONS AND TRANSAXLES JOB SHEET 26

Checking the Transaxle Mounts

Name _____ Station _____ Date _____

NATEF Correlation

This Job Sheet addresses the following **MLR** task:

B.3. Inspect, replace and align powertrain mounts.

This Job Sheet addresses the following **AST/MAST** task:

B.5. Inspect powertrain mounts.

Objective

Upon completion of this job sheet, you will be able to inspect, replace, and align powertrain mounts.

Tools and Materials

Engine support fixture

Engine hoist

Basic hand tools

Protective Clothing

Goggles or safety glasses with side shields

Describe the vehicle being worked on:

Year _____ Make _____ Model _____

VIN _____ Engine type and size _____

Transmission type and model _____

PROCEDURE

1. Many shifting and vibration problems can be caused by worn, loose, or broken engine and transmission mounts. Visually inspect the mounts for looseness and cracks. Give a summary of your visual inspection.

   _____

   _____

   _____

2. Pull up and push down on the transaxle case while watching the mount. If the mount's rubber separates from the metal plate, or if the case moves up but not down, replace the mount. If there is movement between the metal plate and its attaching point on the frame, tighten the attaching bolts to an appropriate torque. Describe the results of doing this.

_____

_____

3. From the driver's seat, apply the foot brake, set the parking brake, and start the engine. Put the transmission into a forward gear and gradually increase the engine speed to about 1500 to 2000 rpm. Watch the torque reaction of the engine on its mounts. If the engine's reaction to the torque appears to be excessive, broken or worn drive train mounts may be the cause. Describe the results of doing this.

_____

_____

4. If it is necessary to replace the transaxle mount, make sure you follow the manufacturer's recommendations for maintaining the alignment of the driveline. Describe the recommended alignment procedures.

_____

_____

_____

_____

5. When removing the transaxle mount, begin by disconnecting the battery's negative cable.

Task Completed ☐

6. Disconnect any electrical connectors that may be located around the mount. Be sure to label any wires you remove to facilitate reassembly.

Task Completed ☐

7. It may be necessary to move some accessories, such as the horn, in order to service the mount without damaging some other assembly.

Task Completed ☐

8. Install the engine support fixture and attach it to an engine hoist.

Task Completed ☐

9. Lift the engine just enough to take the pressure off the mounts.

Task Completed ☐

10. Remove the bolts attaching the transaxle mount to the frame and the mounting bracket, and then remove the mount.

Task Completed ☐

11. To install the new mount, position the transaxle mount in its correct location on the frame and tighten its attaching bolts to the proper torque. What is the torque specification?

_____

12. Install the bolts that attach the mount to the transaxle bracket. Before tightening these bolts, check the alignment of the mount.

Task Completed ☐

13. Once you have confirmed that the alignment is correct, tighten all loosened bolts to their specified torque.

Task Completed ☐

14. Remove the engine hoist fixture from the engine, and reinstall all accessories and wires that may have been removed earlier.

Task Completed ☐

Problems Encountered

_____

_____

_____

Instructor's Comments

_____

_____

_____

AUTOMATIC TRANSMISSIONS AND TRANSAXLES JOB SHEET 27

Prepare to Remove a Transaxle or Transmission

Name _____ Station _____ Date _____

NATEF Correlation

This Job Sheet addresses the following **AST/MAST** task:

C.1. Remove and reinstall transmission/transaxle and torque converter; inspect engine core plugs, rear crankshaft seal, dowel pins, dowel pin holes, and mating surfaces.

Objective

Upon completion of this job sheet, you will be able to describe the procedures that must be followed in order to remove a transaxle or transmission from a vehicle.

Tools and Materials

Hoist   Drain pan

Transmission jack   Droplight or good flashlight

Engine support fixture   Service information

Protective Clothing

Goggles or safety glasses with side shields

Describe the vehicle being worked on:

Year _____ Make _____ Model _____

VIN _____ Engine type and size _____

Model and type of transmission _____

PROCEDURE

1. Refer to the service information and look for any precautions or special instructions that relate to the removal of a transmission from this vehicle. Describe these.

   _____

   _____

   _____

2. Before beginning to remove the transmission from this vehicle, what is the first thing that should be disconnected? Why?

   _____

   _____

3. Look under the hood and identify everything that should be removed or disconnected from above before raising the vehicle on the hoist. If your vehicle is front-wheel drive, do you have to install an engine support fixture before removing the transmission? If so, list these items.

_____

_____

_____

_____

_____

_____

_____

_____

_____

4. Raise the vehicle to a comfortable working height.   Task Completed ☐

5. Carefully examine the area and identify everything that should be disconnected or removed before unbolting the transmission from the engine. List these items.

_____

_____

_____

_____

_____

_____

_____

_____

_____

_____

6. Support the transmission/transaxle with the transmission jack and/or   Task Completed ☐
   holding fixtures.

7. Unbolt the transmission from the engine and slide it away from the engine, making sure the torque converter and oil pump drive is free.

Task Completed ☐

8. Now lower the transmission with the jack and place it in a suitable work area.

Task Completed ☐

9. Carefully inspect the alignment dowel pins and their bores in the engine and the transmission. What did you find?

_____

_____

_____

10. Inspect the area around the crankshaft and flex plate. Is there any evidence of an oil leak from the rear crankshaft seal? If so, what needs to happen?

_____

_____

_____

11. Inspect the area around the crankshaft and flex plate. Is there any evidence of a coolant leak from the engine's core plugs? If so, what needs to happen?

_____

_____

_____

12. Carefully check the mating surfaces of the engine and the transmission/transaxle. Are there any defects that may cause the two units not to mate properly? If so, what needs to be done?

_____

_____

_____

13. Normally, reinstallation is the reverse of the removal procedure. Is this true for this vehicle? If not, what special procedures must be followed?

_____

_____

_____

_____

Problems Encountered

_____

_____

_____

Instructor's Comments

_____

_____

_____

_____

AUTOMATIC TRANSMISSIONS AND TRANSAXLES JOB SHEET 28

Transmission Cooler Inspection and Flushing

Name _____ Station _____ Date _____

NATEF Correlation

This Job Sheet addresses the following **AST/MAST** task:

C.2. Inspect, leak test, and flush or replace transmission/transaxle oil cooler, lines, and fittings.

Objective

Upon completion of this job sheet, you will be able to inspect, leak test, flush, and replace the transmission cooler, lines, and fittings.

Tools and Materials

Line wrenches	Drain pan
Tubing cutter	Compressed air and air nozzle
Tubing flaring kit	Supply of clean solvent or mineral spirits
ATF pressure tester	Lint-free shop towels
Various lengths of rubber hose	Service information

Protective Clothing

Goggles or safety glasses with side shields

Describe the vehicle being worked on:

Year _____ Make _____ Model _____

VIN _____ Engine type and size _____

Model and type of transmission _____

PROCEDURE

1. Describe the condition, smell, and color of the transmission fluid.

   _____

2. If coolant has entered the transmission fluid, what will the fluid look like?

   _____

   _____

3. Open the radiator cap (after the engine is cooled down) and check the coolant for traces of ATF. Also, check the cap and gasket for signs of ATF. Record your findings.

   _____

   _____

4. Replace the radiator cap.

Task Completed ☐

5. Inspect the metal lines and fittings to and from the transmission cooler. Look for damage and signs of leakage. Describe your findings.

_____

_____

6. Summarize the results of your inspection.

_____

_____

7. Place a drain pan under the fittings that connect the cooler lines to the radiator.

Task Completed ☐

8. Disconnect both cooler lines from the radiator. Plug or cap the lines from the transmission.

Task Completed ☐

9. Plug or cap one fitting at the radiator.

Task Completed ☐

10. Remove the radiator cap.

Task Completed ☐

11. Hold the compressed air nozzle tightly against the open fitting.

Task Completed ☐

12. Apply air pressure through the fitting (no more than 75 psi).

Task Completed ☐

13. Check the coolant in the radiator for signs of air bubbles or air movement. Describe your findings.

_____

_____

14. Unplug the other fitting and the cooler lines.

Task Completed ☐

15. Reconnect the cooler lines to the radiator.

Task Completed ☐

16. Summarize the results of the leak test.

_____

_____

_____

_____

Transmission Cooler Flow Test Procedure

Different manufacturers have various procedures to check cooler flow. Some use special tools such as flowmeters to determine if the cooler flow is adequate. The procedure listed is a generic procedure. Look in the service information for your vehicle and perform the recommended procedure if possible.

1. Describe the cooler flow test procedure listed in the service information for your vehicle if applicable.

Task Completed ☐

2. Set the parking brake and start the engine. Allow the engine and transmission to reach normal operating temperature before proceeding. Turn off the engine.

Task Completed ☐

3. Ensure that the transmission fluid level is normal (with or without a dipstick).   Task Completed ☐

4. Raise the vehicle on a hoist.   Task Completed ☐

5. Determine which of the cooler lines is the return line to the transmission. Disconnect the cooler return line at the point where it enters the transmission case.   Task Completed ☐

6. Attach a rubber hose to the disconnected cooler return line.   Task Completed ☐

7. Lower the vehicle and place the end of the hose into a clean container. Clip the line to the container.   Task Completed ☐

8. Start the engine and wait for the air bubbles to clear from the line.   Task Completed ☐

9. Observe the flow of fluid into the container after about **15 seconds**. Describe the flow into the container. What is the specified fluid flow in 15 seconds?   Task Completed ☐

_____

_____

10. Turn off the engine.   Task Completed ☐

11. Raise the vehicle and remove the rubber hose from the return line.   Task Completed ☐

12. Reconnect the return line.   Task Completed ☐

13. Check the transmission fluid level, check for leaks.   Task Completed ☐

   NOTE: *The remainder of this job sheet covers a procedure for flushing the transmission cooler. Check with your instructor before proceeding. Some manufacturers and rebuilders recommend replacing the transmission cooler if the transmission has failed and has left debris in the lines. Some manufacturers have special tools and some use solvents, while others do not.*

1. Describe the recommended procedure to flush or replace the transmission cooler according to the service information. Perform the specific procedure for the vehicle if possible.

_____

_____

_____

2. Raise the vehicle on a hoist.   Task Completed ☐

3. Place a drain pan under the engine's radiator (or place it under the external transmission cooler if the vehicle is so equipped).   Task Completed ☐

4. Disconnect both cooler lines at the radiator or cooler.   Task Completed ☐

5. Clean the fittings at the ends of the lines and at the radiator or cooler.   Task Completed ☐

6. Connect a rubber hose to the inlet fitting at the radiator.   Task Completed ☐

7. Place the other end of the hose into the drain pan.   Task Completed ☐

8. Attach a hose funnel to the cooler return fitting. Install a funnel in the other end of this hose.

Task Completed ☐

9. Pour small amounts of clean transmission fluid into the funnel. Observe the flow and the condition of the fluid moving into the drain pan. What did it look like?

_____

_____

10. If there is little flow, apply some air pressure to the hose connected to the return line.

Task Completed ☐

11. Continue pouring transmission fluid into the funnel until the fluid entering the drain pan is clear.

Task Completed ☐

12. Disconnect the cooler lines at the transmission and place the drain pan at one end of the lines.

Task Completed ☐

13. Install the rubber hose and funnel to the other end of the lines.

Task Completed ☐

14. Pour clean transmission fluid through the lines until the flow is clear. Describe your findings.

_____

_____

15. Pour ATF through the cooler lines to remove any contaminated fluid or solvent residue (if applicable).

Task Completed ☐

16. Remove all hoses and reconnect the cooler lines to the transmission and radiator.

Task Completed ☐

17. Check the transmission's fluid level and correct it if necessary.

Task Completed ☐

Problems Encountered

_____

_____

_____

Instructor's Comments

_____

_____

_____

AUTOMATIC TRANSMISSIONS AND TRANSAXLES JOB SHEET 29

Visual Inspection of a Torque Converter

Name _____ Station _____ Date _____

NATEF Correlation

This Job Sheet addresses the following **AST/MAST** task:

C.3. Inspect converter flex (drive) plate, converter attaching bolts, converter pilot, converter pump drive surfaces, converter end play, and crankshaft pilot bore.

Objective

Upon completion of this job sheet, you will be able to conduct a visual inspection of a torque converter.

Tools and Materials

Clean white paper towels   Droplight or good flashlight

Hoist   Service information

Remote starter switch   Special tools for the torque converter

Dial indicator

Protective Clothing

Goggles or safety glasses with side shields

Describe the vehicle being worked on:

Year _____ Make _____ Model _____

VIN _____ Engine type and size _____

Model and type of transmission _____

PROCEDURE

1. Park the vehicle on a level surface.   Task Completed ☐

2. Describe the condition of the fluid (color, condition, and smell).

    _____

    _____

3. What is indicated by the fluid's condition?

    _____

    _____

    _____

4. Connect a remote starter switch to the vehicle and disable the ignition. Explain how you did this.

_____

_____

5. Raise the vehicle on a hoist to a comfortable working height.　　　　Task Completed ☐

6. Put on eye protection.　　　　Task Completed ☐

7. Remove the torque converter access cover or shield.　　　　Task Completed ☐

8. Inspect the converter through the access hole. Use the remote starter switch to observe the entire converter. Describe your findings.

_____

_____

9. Carefully look for signs of fluid leakage around the torque converter and the access cover or shield. Describe your findings and state what could be causing the leak.

_____

_____

10. Look at the heads of all torque converter bolts. Do they show signs of contact with other parts?

_____

_____

Out-of-Vehicle Torque Converter Checks

1. Detailed diagnosis of a torque converter takes place with the torque converter removed from the vehicle. Inspect the drive studs or lugs used to attach the converter to the flex-plate. Summarize your findings.　　　Task Completed ☐

2. Check the threads of the lugs and studs. If they are damaged slightly, they can be cleaned up with a thread file, tap, or die set. However, if they are badly damaged, the converter should be replaced. The converter should also be replaced if the studs or lugs are loose or damaged. Carefully check the weld spots that position the studs and lugs on the converter. If the welds are cracked, replace the converter. Summarize your findings.　　　Task Completed ☐

3. Check the hub of the converter for scoring, nicks, excessive wear, burrs, and signs of discoloration. If polishing cannot remove the defects in the hub, the converter should be replaced. Summarize your findings.　　　Task Completed ☐

4. Carefully check the converter's flex plate for damage and cracks. Also check the ring gear for looseness. Summarize your findings.　　　Task Completed ☐

5. Noises may be caused by internal converter parts hitting each other or hitting the housing. To check for any interference between the stator and turbine, place the converter, face down, on a bench.　　　Task Completed ☐

6. Install the oil pump assembly. Make sure the oil pump drive engages with the oil pump.

Task Completed ☐

7. Insert the input shaft into the hub of the turbine.

Task Completed ☐

8. Hold the oil pump and converter stationary, and then rotate the turbine shaft in both directions. If the shaft does not move freely and/or makes noise, the converter must be replaced. Summarize your findings.

_____

_____

9. To check for any interference between the stator and the impeller, place the transmission's oil pump on a bench and fit the converter over the stator support splines. Rotate the converter until the hub engages with the oil pump drive. Then, hold the pump stationary and rotate the converter in a counterclockwise direction. If the converter does not freely rotate or makes a scraping noise during rotation, the converter must be replaced. Summarize your findings.

_____

_____

10. Check end play with the proper holding tool and a dial indicator with a holding fixture. Insert the holding tool into the hub of the converter and once bottomed, tighten it in place.

Task Completed ☐

11. This locks the tool into the splines of the turbine while the dial indicator is fixed onto the hub. The amount indicated on the dial indicator, as the tool is lifted up, is the amount of end play inside the converter. If this amount exceeds specifications, replace the converter. Summarize your findings.

_____

_____

12. If the initial visual inspection suggested that the converter has a leak, special test equipment can be used to determine if the converter is leaking. This equipment uses compressed air to pressurize the converter. Leaks are found in much the same way as tire leaks are—that is, the converter is submerged in water and the trail of air bubbles leads the technician to the source of leakage. This test equipment can only be used to check converters with a drain plug and therefore is somewhat limited on current applications. However, some manufacturers list a procedure for installing a drain plug or draining the converter in their service information. Summarize your findings.

_____

_____

Problems Encountered

_____

_____

_____

Instructor's Comments

_____

_____

_____

AUTOMATIC TRANSMISSIONS AND TRANSAXLES JOB SHEET 30

Continuously Variable Transmissions

Name _____ Station _____ Date _____

NATEF Correlation

This Job Sheet addresses the following **MLR** task:

C.1. Describe the operational characteristics of a continuously variable transmission (CVT).

This Job Sheet addresses the following **AST/MAST** task:

C.4. Describe the operational characteristics of a continuously variable transmission (CVT).

Objective

Upon completion of this job sheet, you will be able to explain how a normal CVT works.

Tools and Materials

Service Information

Describe the vehicle being worked on:

Year _____ Make _____ Model _____

VIN _____ Engine type and size _____

Transmission type and model _____

PROCEDURE

1. Why is this type of transmission called a continuously variable transmission?

   _____

   _____

2. What is the basic goal of a CVT?

   _____

   _____

3. Describe the major components of this transmission.

   _____

   _____

   _____

   _____

4. Explain how this CVT works.

_____

_____

_____

_____

5. Does the driver feel a change in ratio as the vehicle is driven? Why or why not?

_____

_____

_____

_____

6. Using the service information as a guide, what inputs does the transmission rely on to change ratios?

_____

_____

_____

_____

7. Explain how reverse is achieved in this transmission.

_____

_____

_____

_____

Problems Encountered

_____

_____

_____

Instructor's Comments

_____

_____

_____

AUTOMATIC TRANSMISSIONS AND TRANSAXLES JOB SHEET 31

Hybrid Transmissions

Name _____ Station _____ Date _____

NATEF Correlation

This Job Sheet addresses the following **MLR** task:

C.2. Describe the operational characteristics of a hybrid vehicle drive train.

This Job Sheet addresses the following **AST/MAST** task:

C.5. Describe the operational characteristics of a hybrid vehicle drive train.

Objective

Upon completion of this job sheet, you will be able to explain how a transmission in a hybrid vehicle works.

Tools and Materials

Service information

Describe the vehicle being worked on:

Year _____ Make _____ Model _____

VIN _____ Engine type and size _____

PROCEDURE

1. Using the service information, briefly describe the hybrid system used on this vehicle.

   _____

   _____

   _____

   _____

2. Describe the key characteristics that make this a hybrid vehicle.

   _____

   _____

   _____

   _____

3. Where is the electric motor(s) located?

_____

_____

_____

_____

4. What are principle purposes of the electric motor?

_____

_____

_____

_____

5. What is the voltage rating of the hybrid battery?

_____

_____

6. Does this vehicle use a conventional transmission?

_____

_____

7. If the vehicle has a unique transmission, describe why it is unique.

_____

_____

_____

_____

8. Can this vehicle run on electric power only? If so, when?

_____

_____

_____

_____

9. Briefly explain the action of the electric motor(s). When are they turned on and when do they function as a generator?

_____

_____

10. Briefly describe the precautions that must be followed when servicing these transmissions and vehicles.

_____

_____

_____

_____

Problems Encountered

_____

_____

_____

Instructor's Comments

_____

_____

_____

AUTOMATIC TRANSMISSIONS AND TRANSAXLES JOB SHEET 32

Disassemble and Inspect a Transmission

Name _____ Station _____ Date _____

NATEF Correlation

This Job Sheet addresses the following **MAST** task:

C.6. Disassemble, clean, and inspect transmission/transaxle.

Objective

Upon completion of this job sheet, you will be able to disassemble and inspect the major components of a transmission.

Tools and Materials

Compressed air and air nozzle   Lint-free shop towels

Supply of clean solvent   Service information

Protective Clothing

Goggles or safety glasses with side shields

Describe the vehicle being worked on:

Year _____ Make _____ Model _____

VIN _____ Engine type and size _____

Model and type of transmission _____

PROCEDURE

1. Disassemble the transmission into major units. Set each unit aside until this job sheet refers to it. Describe any problems you encountered while disassembling the transmission. Be sure to follow the procedures given in the service information while taking the transmission apart.

   _____

   _____

   _____

   _____

2. Clean the planetary gear set in clean solvent and allow it to air dry. If compressed air is used to help the drying process, firmly hold the pinion gears to prevent the bearings from moving. What did you use to clean them?

   _____

   _____

3. Clean all thrust washers, thrust bearings, and bushings in clean solvent and allow them to air dry. Then, carefully inspect them. What did you find?

_____

_____

_____

_____

4. Place one member of the planetary gear set on the sun gear. Rotate the gear set slowly. Do the same for the other parts of the gear set. Describe the feel of the gears' rotation.

_____

_____

_____

_____

5. Inspect each member of the gear set for damaged or worn gear teeth. Describe their condition.

_____

_____

6. Inspect the splines of each gear set member and describe their condition.

_____

_____

7. Remove any buildup of material or dirt that may be present between the teeth of the gears. How did you do this?

_____

_____

8. Look at the carrier assemblies and check for cracks or other damage. Describe your findings.

_____

_____

9. Inspect the entire gear set for signs of discoloration. Describe your findings and explain what is indicated by them.

_____

_____

10. Check the output shaft and its splines for wear, cracking, or other damage. Describe their condition.

   _____

   _____

11. Carefully inspect the driving shells and drive lugs for wear and damage. Describe their condition.

   _____

   _____

12. Summarize the condition of the planetary gear sets and shaft.

   _____

   _____

   _____

   _____

13. Check the thrust washers and bearings for damage, excessive wear, and distortion. Describe their condition.

   _____

   _____

14. Check all of the bushings for signs of wear, scoring, and other damage. Describe their condition.

   _____

   _____

   _____

   _____

15. Check the shafts that ride in the bushings. Describe their condition.

   _____

   _____

16. Summarize the condition of the thrust washers, bearings, and their associated shafts.

   _____

   _____

   _____

   _____

Problems Encountered

_____

_____

_____

Instructor's Comments

_____

_____

_____

AUTOMATIC TRANSMISSIONS AND TRANSAXLES JOB SHEET 33

Service a Valve Body

Name _____ Station _____ Date _____

NATEF Correlation

This Job Sheet addresses the following **MAST** task:

C.7. Inspect, measure, clean, and replace valve body (includes surfaces, bores, springs, valves, sleeves, retainers, brackets, check valves/balls, screens, spacers, and gaskets).

Objective

Upon completion of this job sheet, you will be able to demonstrate the ability to disassemble, clean, inspect, and reassemble a valve body.

Tools and Materials

Transmission stand or holding fixture   Cleaning solvent

Compressed air and nozzle   Lint-free shop rags

Pound-inch torque wrench   Service information

Measuring calipers

Protective Clothing

Goggles or safety glasses with side shields

Describe the vehicle being worked on:

Model and type of transmission _____

Vehicle the transmission is from _____

Year _____ Make _____ Model _____

VIN _____ Engine type and size _____

PROCEDURE

1. Mount the transmission to the stand or holding fixture with the valve body facing up.   Task Completed ☐

2. Remove the valve body attaching screws. Start at the outside bolts and work toward the center if the service information does not give specific directions.   Task Completed ☐

3. Remove the valve body assembly. Remove all the electrical solenoids and wiring and place them in a container of clean solvent.   Task Completed ☐

4. Remove the end plates and covers from the assembly.   Task Completed ☐

5. Below, draw out a simplified view of the valve body. Note the location of all valves, check balls, and springs.

6. Begin to disassemble the unit. Lay all of the parts on a clean surface in the order in which they were removed. This will help during reassembly. Be careful with the check balls, since they may be different sizes. Some transmissions use slightly larger or smaller check balls in circuits. How many check balls were in the valve body? _____ Were they all the same size? _____

7. Remove all of the gaskets from the assembly. Place them aside. Do not throw them away. They will be needed for comparison when installing new gaskets.   Task Completed ☐

8. Thoroughly clean the main body and plates of the assembly. Do not dry the valve body with anything that will leave lint. Allow the unit to air dry or dry it off with a gentle flow of compressed air.   Task Completed ☐

9. Inspect the valve body for damage and cracks. Also check the flatness of the body and the plates. Describe your findings.

   _____

   _____

   _____

10. What are the valves made of ? _____

11. Check all the valves for free movement in their bores. Also check the valves for wear, scoring, and signs of sticking. Describe your findings and recommendations.

    _____

    _____

    _____

12. Correctly install the valves and springs in their bores.   Task Completed ☐

13. Replace the end plates or covers. Install the retaining screws by hand, and then tighten them to the specific torque. The specified torque is _____.

14. Test and install the solenoids removed from the valve body previously.   Task Completed ☐

15. Install the check balls into their proper location.   Task Completed ☐

16. Compare the new gaskets with the old and install the correct ones on the valve body.

Task Completed ☐

17. Align the spring-loaded check balls and all other parts as needed.

Task Completed ☐

18. Place the gasket on top of the transfer plate. Align the gasket holes with the transfer plate and the bores in the valve body.

Task Completed ☐

19. Position the valve body on the transmission. Align the parking and other internal linkages. Then, install the retaining screws by hand.

Task Completed ☐

20. Tighten the screws to the specified torque and in the specified order. The specified torque is _____.

Problems Encountered

_____

_____

_____

Instructor's Comments

_____

_____

_____

AUTOMATIC TRANSMISSIONS AND TRANSAXLES JOB SHEET 34

Servo and Accumulator Service

Name _____ Station _____ Date _____

NATEF Correlation

This Job Sheet addresses the following **MAST** task:

C.8. Inspect servo and accumulator bores, pistons, seals, pins, springs, and retainers; determine necessary action.

Objective

Upon completion of this job sheet, you will be able to inspect the servo's and accumulator's bore, piston, seals, pin, spring, and retainers.

Tools and Materials

Snap ring pliers   Small file

Crocus cloth   Knife

Protective Clothing

Goggles or safety glasses with side shields

Describe the vehicle being worked on:

Year _____ Make _____ Model _____

VIN _____ Engine type and size _____

Transmission type and model _____

PROCEDURE

1. On some transmissions, the servo and accumulator assemblies are serviceable with the transmission in the vehicle. Others require the complete disassembly of the transmission. Before disassembling a servo or any other component, carefully inspect the area to determine the exact cause of the leakage. Do this before cleaning the area around the seal. Look at the path of the fluid leakage and identify other possible sources. These sources could be worn gaskets, loose bolts, cracked housings, or loose line connections. Describe your findings.

   _____

   _____

2. Inspect the outside area of the seal. If it is wet, determine if the oil is leaking out or if it is merely a lubricating film of oil. Describe your findings.

_____

_____

3. When removing the servo, continue to look for causes of the leak. Check both the inner and outer parts of the seal for wet oil, which means leakage. Describe your findings.

_____

_____

4. When removing the seal, inspect the sealing surface, or lips. Look for unusual wear, warping, cuts and gouges, or particles embedded in the seal. Describe your findings.

_____

_____

5. Remove the retaining rings and pull the assembly from the bore for cleaning. Carefully check the bore. What is its condition?

_____

_____

6. Check the condition of the piston and springs. Cast-iron seal rings may not need replacement, but rubber and elastomer seals should always be replaced. Describe your findings.

_____

_____

7. Begin the disassembling of an accumulator by removing the accumulator plate snap ring.

Task Completed ☐

8. Remove the accumulator plate, the spring, and the accumulator pistons. If rubber seal rings are installed on the piston, replace them whenever you are servicing the accumulator.

Task Completed ☐

9. Lubricate the new accumulator piston ring and carefully install it on the piston.

Task Completed ☐

10. Lubricate the accumulator cylinder walls and install the accumulator piston and spring.

Task Completed ☐

11. Install the accumulator plate and retaining snap ring.

Task Completed ☐

12. A servo is serviced in a similar fashion. The servo's piston, spring, piston rod, and guide should be cleaned and dried.

Task Completed ☐

13. Check the servo piston for cracks, burrs, scores, and wear. Servo pistons should be carefully checked for cracks and their fit on the guide pins. Describe your findings.

_____

_____

14. Check the piston's bore and seal groove for defects or damage. Describe your findings.

_____

_____

15. Check cast-iron seal rings to make sure they are able to turn freely in the piston groove. These seal rings are not typically replaced unless they are damaged, so carefully inspect them. Describe your findings.

_____

_____

16. Inspect the servo or accumulator spring for possible cracks. Also check where the spring rests against the case or piston. Describe your findings.

_____

_____

17. Inspect the servo cylinder for scores or other damage. Describe your findings.

_____

_____

18. Move the piston rod through the piston rod guide and check for freedom of movement. Describe your findings.

_____

_____

19. Check the band servo components for wear and scoring. Describe your findings.

_____

_____

20. When reassembling the servo, lubricate the seal ring with ATF and carefully install it on the piston rod.   Task Completed ☐

21. Lubricate and install the piston rod guide with its snap ring into the servo piston.   Task Completed ☐

22. Install the servo piston assembly, return spring, and piston guide into the servo cylinder.   Task Completed ☐

23. Lubricate and install the new lip seal.   Task Completed ☐

Problems Encountered

_____

_____

_____

Instructor's Comments

_____

_____

_____

AUTOMATIC TRANSMISSIONS AND TRANSAXLES JOB SHEET 35

Reassembly of a Transmission/Transaxle

Name _____ Station _____ Date _____

NATEF Correlation

This Job Sheet addresses the following **MAST** tasks:

C.9. Assemble transmission/transaxle.

C.11. Measure transmission/transaxle end play or preload; determine necessary action.

Objective

Upon completion of this job sheet, you will be able to reassemble an automatic transmission and properly measure and set end play and preload during the assembly of a transmission or transaxle.

Tools and Materials

Air nozzle

Dial indicator

Service information

Cleaning ATF in a tray

Petroleum jelly or assembly lube

Protective Clothing

Goggles or safety glasses with side shields

Describe the vehicle being worked on:

Year _____ Make _____ Model _____

VIN _____ Engine type and size _____

Transmission type and model _____

PROCEDURE

1. Before proceeding with the final assembly of all components, it is important to verify that the case, housing, and parts are clean and free from dust, dirt, and foreign matter.

   Task Completed ☐

2. Coat all parts just before installation with the proper type of ATF. Soak bands and clutches in the fluid for at least 15 minutes before installing them. All new seals and rings should have been installed before beginning final assembly.

   Task Completed ☐

3. Examine all thrust washers carefully and coat them with petroleum jelly or transmission assembly gel before placing them in the housing. Follow the assembly process carefully. Remember that all moving parts are separated with a thrust washer or bearing assembly.

   Task Completed ☐

4. Start assembly by referring to the service information for the vehicle. Where do you start your assembly?

_____

_____

_____

5. Assemble the various clutch packs as needed. Replace the piston seals, paying close attention to the way the lips of the seals are facing. Sometimes a special tool is used to reassemble the seals to the piston housings. Describe the process of assembling the piston seals and pistons into the clutch packs.

_____

_____

_____

6. After the clutch piston has been installed, the return springs are installed, generally with the aid of a special tool. Generally, a snap ring is installed to hold the return springs in place. Describe the process of installing the return springs on one of your clutch pack assemblies.

_____

_____

_____

7. After determining the correct number of friction and steel plates, install the steel dished plate first, then the steel and friction plates, and finally the retaining plate and snap ring, all according to the service information. How many steel plates did you have?

_____

8. After the clutch pack has been completely assembled, measure the clearance between the snap ring and the retainer plate. Select the proper thickness of retaining plate that will give the correct ring-to-plate clearance if the measurement does not meet the specified limits according to the service information. What were the results?

_____

9. Using a suitable blowgun with a rubber tapered tip, air-check the clutch operation. What were the results?

_____

_____

10. Continue to replace the clutch packs until they have all been serviced. How many clutch packs did you service?

_____

11. Start the buildup of the transmission according to the service information. Be sure to make note of each part as it is installed. Did you have any problems with the assembly?

_____

_____

_____

12. End-play measurements are a critical part of any transmission rebuild. How many end-play measurements are you required to perform on your assigned transmission?

_____

_____

_____

13. Why are end-play measurements important?

_____

_____

> **NOTE:** _Before air-checking any component, make sure you are wearing eye protection. Do not air-check a piston until the clutch pack has the clutches and snap rings installed. The clutch piston might be overextended if it is not restrained by the clutch assembly. Also, never use more air pressure than is recommended by the manufacturer, so as to prevent any damage or personal injury._

14. Air-check the transmission assembly before completing the rebuild process, generally before installing the valve body. Does the service information recommend air-checking the transmission before final assembly?

_____

_____

_____

15. Describe the number of air-checks recommended by the manufacturer.

_____

_____

16. Once the transmission is reassembled, the end play of the transmission assembly must be checked. Install a dial indicator and check the travel of the output shaft as it is pushed and pulled. What were the results?

_____

_____

17. Were there any other end-play measurements recommended during the overhaul process?

_____

_____

_____

18. Why is end play an important factor in a transmission rebuild?

_____

_____

19. To assemble the parking pawl assembly place the assembly into its position and install the extension housing with a new gasket, then tighten the attaching bolts to the proper specifications. What are the specifications?

_____

20. On transaxles, the differential assembly should be disassembled, cleaned, inspected, and reassembled. After it has been reassembled, measure its end play with gauging shims. Then, select a shim thick enough to correct the end play. What are the specifications?

_____

_____

_____

21. Install the valve body. Be sure the manual valve is in alignment with the selector pin. Tighten the valve body attaching bolts to the specified torque. What are the specifications?

_____

_____

_____

22. Before installing the oil pan, check the alignment and operation of the control lever and parking pawl engagement. Make a final check to be sure all bolts are installed in the valve body.   Task Completed ☐

23. Install the oil pan with a new gasket. Torque the bolts to specifications. What are the specifications?

_____

24. Lubricate the oil pump's lip seal and the converter neck before installing the converter.   Task Completed ☐

25. Install the converter, making sure the converter is properly meshed with the oil pump drive gear.   Task Completed ☐

26. The transmission is now ready for installation into the vehicle. Use the reverse of the removal procedures. Remember to follow proper fluid filling procedures.   Task Completed ☐

Problems Encountered

_____

_____

_____

Instructor's Comments

_____

_____

_____

Problems Encountered

Instructor's Comments

AUTOMATIC TRANSMISSIONS AND TRANSAXLES JOB SHEET 36

Servicing an Automatic Transmission Oil Pump

Name _____ Station _____ Date _____

NATEF Correlation

This Job Sheet addresses the following **MAST** task:

C.10. Inspect, measure, and reseal oil pump assembly and components.

Objective

Upon completion of this job sheet, you will be able to inspect, measure, and replace the components of an oil pump assembly. Both vane and gear styles are included in the job sheet.

Tools and Materials

Machinist dye

Feeler gauge

Protective Clothing

Goggles or safety glasses with side shields

Describe the vehicle being worked on:

Year _____ Make _____ Model _____

VIN _____ Engine type and size _____

PROCEDURE

Vane-style Oil Pumps

1. Loosen the assembly hold-down bolts evenly to remove the pump. (Some rear-wheel drive pumps require a special tool to remove them from the case.)

   Task Completed ☐

2. Pay particular attention to the mounting bolts, since several different lengths can be involved. If possible, remove the oil pump from the transmission as an assembly, to prevent the vanes from falling out of the pump. Describe the process of removing the oil pump.

   _____

   _____

   _____

3. Using the service information as a guide, separate the vane pump halves. Carefully inspect the housings for any scoring or nicks that may cause internal leakage. Describe your findings.

_____

_____

_____

4. Check the service information for any measurements that need to be made to the pump internally. List your findings.

_____

_____

_____

5. Describe the general condition of the oil pump.

_____

_____

_____

6. Replace the internal seals as necessary and carefully reassemble the vanes back into the pump. Describe the seals used in the pump and the process you used to replace them.

_____

_____

PROCEDURE

Gear-style Oil Pumps

1. With the oil pump removed from the transmission, remove the front pump bearing race, front clutch thrust washer, gasket, and O-ring. Inspect the pump bodies, pump shaft, and ring groove areas. Record your findings.

_____

_____

_____

2. Mount the pump on a stand and unbolt and separate the pump bodies.   Task Completed ☐

3. Mark the gears with machinist bluing ink or paint before removing them so that the gears will remain in the same relationship during reassembly. What did you use to mark them and where did you mark them?

_____

_____

4. Inspect the gears and all internal surfaces for defects and visible wear. Record your findings.

_____

_____

5. With the pump mounted on the stand, use a feeler gauge to measure between the outer gear and the crest in the pump housing crest. What tolerance do the specifications call for? What did you measure?

_____

_____

6. Measure between the outer gear teeth and the crescent. What tolerance do the specifications call for? What did you measure?

_____

_____

7. Place the pump flat on the bench, and with a feeler gauge and straightedge measure between the gears and the pump cover. What tolerance do the specifications call for? What did you measure?

_____

_____

8. Measure the clearance between the C-ring and the ring groove. What tolerance do the specifications call for? What did you measure?

_____

_____

9. Using the stand or a centering band to center the pump, torque the securing bolts to specifications. What is the specified torque?

_____

_____

10. Replace all O-rings and gaskets.   Task Completed ☐

Problems Encountered

_____

_____

_____

Instructor's Comments

_____

_____

_____

AUTOMATIC TRANSMISSIONS AND TRANSAXLES JOB SHEET 37

Checking Thrust Washers, Bushings, and Bearings

Name _____ Station _____ Date _____

NATEF Correlation

This Job Sheet addresses the following **MAST** tasks:

> **C.12.** Inspect, measure, and replace thrust washers and bearings.
>
> **C.14.** Inspect bushings; determine necessary action.

Objective

Upon completion of this job sheet, you will be able to inspect, measure, and replace thrust washers and bearings and inspect the bushings in a transmission/transaxle.

Tools and Materials

Wire-type feeler gauge set

Bushing puller tool

Bushing driver set

Protective Clothing

Goggles or safety glasses with side shields

Describe the vehicle being worked on:

Year _____ Make _____ Model _____

VIN _____ Engine type and size _____

PROCEDURE

1. The best time to inspect thrust washers, bearings, and bushings is during disassembly. The bushings should be inspected for pitting and scoring. Describe their condition.

   _____

   _____

2. Check the depth to which the bushings are installed and the direction of the oil groove, if so equipped, before you remove them. Many bushings used in the planetary gearing and output shaft areas have oiling holes in them. Be sure to line the oiling holes up correctly during installation or you may block off oil delivery and destroy the gear train. Describe their condition.

   _____

   _____

3. Observe the lateral movement of the shaft that fits into the bushing. Any noticeable lateral movement indicates wear and the bushing should be replaced. Describe your findings.

   _____

   _____

4. The amount of clearance between the shaft and the bushing can be checked with a wire-type feeler gauge. Insert the wire between the shaft and the bushing. If the gap is greater than the maximum allowable, the bushing should be replaced. What are the specifications for this gap and how do they compare to your measurement?

   _____

   _____

5. Measure the inside diameter of the bushing and the outside diameter of the shaft with a vernier-type caliper or micrometer. Compare the two and state your conclusions.

   _____

   _____

6. Most bushings are press-fit into a bore. To remove them, they are driven out of the bore with a properly sized bushing tool. Some bushings can be removed with a slide hammer fitted with an expanding or threaded fixture that grips to the inside of the bushing. Another way to remove bushings is to carefully cut one side of the bushing and collapse it. Once collapsed, the bushing can be easily removed with a pair of pliers. What did you use to remove the bushing?

   _____

   _____

7. Small-bore bushings located in areas where it is difficult to use a bushing tool can be removed by tapping the inside bore of the bushing with threads that match a bolt that fits into the bushing. After the bushing has been tapped, insert the bolt and use a slide hammer to pull the bolt and bushing out of its bore.

   Task Completed ☐

8. All new bushings should be pre-lubed during transmission assembly and installed with the proper bushing driver. Make sure they are not damaged and are fully seated in their bore.

   Task Completed ☐

9. The purpose of a thrust washer is to support a thrust load and keep parts from rubbing together. Selective thrust washers come in various thicknesses to take up clearances and adjust shaft end play. Flat thrust washers and bearings should be inspected for scoring, flaking, and wear that goes through to the base material. Describe their condition.

_____

_____

10. Flat thrust washers should also be checked for broken or weak tabs. These tabs are critical for holding the washer in place. On metal type flat thrust washers, the tabs may appear cracked at the bend of the tab; however, this is a normal appearance. Describe their condition.

_____

_____

11. Only damaged plastic thrust washers will show wear. The only way to check their wear is to measure the thickness and compare it to a new part. Describe their condition.

_____

_____

12. Proper thrust washer thicknesses are important to the operation of an automatic transmission. Always follow the recommended procedure for selecting the proper thrust plate. Use petroleum jelly or transjel to hold the thrust washers in place during assembly.

Task Completed ☐

CAUTION: _Never use white lube or chassis lube. These greases will not mix with the fluid and can plug up orifices and passages, and hold check balls off their seats._

13. All bearings should be checked for roughness before and after cleaning.

Task Completed ☐

14. Carefully examine the inner and outer races, as well as the rollers, needles, and balls for cracks, pitting, etching, or signs of overheating. Describe their condition.

_____

_____

15. Give a summary of your inspection.

_____

_____

Problems Encountered

_____

_____

_____

Instructor's Comments

_____

_____

_____

AUTOMATIC TRANSMISSIONS AND TRANSAXLES JOB SHEET 38

Servicing Oil Delivery Seals

Name _____ Station _____ Date _____

NATEF Correlation

This Job Sheet addresses the following **MAST** task:

C.13. Inspect oil delivery circuits, including seal rings, ring grooves, and sealing surface areas, feed pipes, orifices, and check valves/balls.

Objective

Upon completion of this job sheet, you will be able to inspect oil delivery seal rings, ring grooves, and sealing surface areas.

Tools and Materials

Petroleum jelly   Crocus cloth

Seal driver tools   Feeler gauge set

Protective Clothing

Goggles or safety glasses with side shields

Describe the vehicle being worked on:

Year _____ Make _____ Model _____

VIN _____ Engine type and size _____

Transmission type and model _____

GUIDELINES

- Three types of seals are used in automatic transmissions: O-ring and square-cut (lathe-cut), lip, and sealing rings. These seals are designed to stop fluid from leaking out of the transmission and to stop fluid from moving into another circuit of the hydraulic circuit.

- O-ring and square-cut seals are used to seal non-rotating parts. When installing a new O-ring or square-cut seal, coat the entire surface of the seal with assembly lube or petroleum jelly. Make sure you don't stretch or distort the seal while you work it into its holding groove. After a square-cut seal is installed, double check it to make sure it is not twisted. The flat surface of the seal should be parallel with the bore. If it is not, fluid will easily leak past the seal.

- Lip seals that are used to seal a shaft typically have a metal flange around their outside diameter. The shaft rides on the lip seal at the inside diameter of the seal assembly. The rigid outer diameter provides a mounting point for the lip seal and is pressed into a bore. Once pressed into the bore, the outer diameter of the seal prevents fluid from leaking into the bore, while the inner lip seal prevents leakage past the shaft.

- Piston lip seals are set into a machined groove on the piston. This type of lip seal is not housed in a rigid metal flange. They are designed to be flexible and provide a seal while the piston moves up and down. While the piston moves, the lip flexes up and down. The most important thing to keep in mind while installing a lip seal is to make sure the lip is facing in the correct direction. The lip should always be aimed toward the source of the pressurized fluid. If installed backwards, fluid under pressure will easily leak past the seal. Also remember to make sure the surfaces to be sealed are clean and not damaged.

- Teflon or metal sealing rings are commonly used to seal servo pistons, oil pump covers, and shafts. These rings may be designed to provide for a seal, but they may also be designed to allow a controlled amount of fluid leakage. Sealing rings are either solid rings or are cut. Cut sealing rings are one of three designs: open-end, butt-end, or locking-end.

- Solid sealing rings are made of a Teflon-based material and are never reused. To remove them, simply (but carefully) cut the seal after it has been pried out of its groove. Installing a new solid sealing ring requires special tools. These tools allow you to stretch the seal while pushing it into position. Never attempt to install a solid seal without the proper tools. Because these seals are soft, they are easily distorted and damaged.

- Open-end sealing rings fit loosely into a machined groove. The ends of the rings do not touch when they are installed. This type of ring is typically removed and installed with a pair of snap ring pliers. The ring should be expanded just enough to move it off or onto the shaft.

- Butt-end sealing rings are designed so that their ends butt up or touch each other once the seal is in place. This type of seal can be removed with a small screwdriver. The blade of the screwdriver is used to work the ring out of its groove. To install this type of ring, use a pair of snap ring pliers and expand the ring to move it into position.

- Locking-end rings may have hooked ends that connect or have ends that are cut at an angle to hold the ends together. These seals are removed and installed in the same way as butt-end rings. After these rings are installed, make sure the ends are properly positioned and touching.

- All seals should be checked in their own bores prior to installation. They should be slightly smaller or larger (1 or – 3%) than their groove or bore. If a seal is not the proper size, find one that is. Do not assume that because a particular seal came with the overhaul kit it is the correct one.

- Never install a seal when it is dry. The seal should slide into position and allow the part it seals to slide into it. A dry seal is easily damaged during installation.

- Install only genuine seals recommended by the manufacturer of the transmission.

PROCEDURE

1. The transmission case should be thoroughly cleaned and all passages and orifice tubes blown out.

   Task Completed ☐

2. After the case has been cleaned, all bushings, fluid passages, bolt threads, clutch plate splines, and bores should be checked. The passages can be checked for restrictions and leaks by applying compressed air to each one. If the air comes out the other end, there is no restriction. To check for leaks, plug off one end of the passage and apply air to the other. If pressure builds in that passage, there are probably no leaks in it. What did you find?

   _____

   _____

3. Passages in the case guide the flow of fluid through the case. Although not that common, porosity in this area can cause cross-tracking of one circuit to another. This can cause bind up (two gears at once) or a slow pressure bleed off in the affected circuit. Did the transmission exhibit this type of problem?

   _____

   _____

4. If this is suspected, try filling the circuit with solvent and watching to see if the solvent disappears or leaks away. If a leak is suspected, check each part of the circuit to find where the leak is. Describe your findings.

   _____

   _____

5. The small screens found during teardown should be inspected for foreign material. These screens are used to prevent valve hang up at the pressure regulator and governor and must be in place. Most screens can be removed easily. Care should be taken when cleaning since some cleaning solvents will destroy the plastic screens. Low air pressure (approximately 30 psi) can be used to blow the screens out in a reverse direction. What was the condition of the screens?

   _____

   _____

6. Carefully inspect these areas for damage. File or stone away any burrs or bad nicks and polish the surfaces with a fine crocus cloth. Then, clean the area to remove the metal particles. Describe your findings.

   _____

   _____

7. Make sure the oil passage to a pressure-fed bushing or bearing is open and free of dirt and foreign material. It does no good to replace a bushing without making sure there is good oil flow. Describe your findings.

_____

_____

8. Carefully inspect these areas for damage. File or stone away any burrs or bad nicks and polish the surfaces with a fine crocus cloth. Then, clean the area to remove the metal particles. Describe your findings.

_____

_____

9. Vents are located in the pump body or transmission case and provide for equalization of pressures in the transmission. These vents can be checked by blowing low-pressure air through them, squirting solvent or brake cleaning spray through them, or by pushing a small diameter wire through the vent passage. A clean, open passage is all you need to verify proper operation. Describe your findings.

_____

_____

10. Before installing seals, clean the shaft and/or bore area.   Task Completed ☐

11. Carefully inspect these areas for damage. File or stone away any burrs or bad nicks and polish the surfaces with a fine crocus cloth. Then, clean the area to remove the metal particles. Describe your findings.

_____

_____

12. Lubricate the seal, especially any lip seals, to ease installation.   Task Completed ☐

13. All metal sealing rings should also be checked for proper fit. Since these rings seal on their outer diameter, the seal should be inserted in its bore and should feel tight there. If the seal has some form of locking ends, these should be interlocked prior to trying the seal in its bore. Describe your findings.

_____

_____

14. Check the fit of the sealing rings in their shaft groove. Describe your findings.

_____

_____

15. Check the side clearance of the ring by placing the ring into its groove and measuring the clearance between the ring and the groove with a feeler gauge. Describe your findings.

_____

_____

16. While checking the clearance, look for nicks in the grooves and for evidence of groove taper or stepping. Describe your findings.

_____

_____

17. Use the correct driver when installing a seal and be careful not to damage the seal during installation.   Task Completed ☐

Problems Encountered

_____

_____

_____

Instructor's Comments

_____

_____

_____

AUTOMATIC TRANSMISSIONS AND TRANSAXLES JOB SHEET 39

Servicing Planetary Gear Assemblies

Name _____ Station _____ Date _____

NATEF Correlation

This Job Sheet addresses the following **MAST** task:

C.15. Inspect and measure planetary gear assembly components; determine necessary action.

Objective

Upon completion of this job sheet, you will be able to inspect and measure planetary gear assemblies.

Tools and Materials

Snap ring pliers

Feeler gauge set

Protective Clothing

Goggles or safety glasses with side shields

Describe the vehicle being worked on:

Year _____ Make _____ Model _____

VIN _____ Engine type and size _____

Transaxle type and model _____

PROCEDURE

1. The planetary gears used in automatic transmissions are the helical-type gear and all gear teeth should be inspected for chips or stripped teeth. Describe their general condition before disassembling the gear set.

   _____

   _____

2. Any gear that is mounted to a splined shaft needs the splines checked for mutilation or shifted splines. Record your findings.

   _____

   _____

3. Note any discoloration of the parts and explain the cause for it.

   _____

   _____

4. Check the planetary pinion gears for loose bearings. Record your findings.

_____

_____

5. Check each gear individually by rolling it on its shaft to feel for roughness or binding of the needle bearings. Wiggle the gear to be sure it is not loose on the shaft. Looseness will cause the gear to whine when it is loaded. Record your findings.

_____

_____

6. Inspect the gears' teeth for chips or imperfections as these will also cause whine. Record your findings.

_____

_____

7. Check the gear teeth around the inside of the front planetary ring gear. Record your findings.

_____

_____

8. Check the fit between the front planetary carrier to the output shaft splines. Record your findings.

_____

_____

9. Remove the snap ring and thrust washer from the front planetary ring gear. Record your findings.

_____

_____

10. Examine the thrust washer and the outer splines of the front drum for burrs and distortion. Record your findings.

_____

_____

11. With the snap ring removed, the front planetary carrier can be removed from the ring gear. Check the planetary carrier gears for end play by placing a feeler gauge between the planetary carrier and the planetary pinion gear. Compare the end play to specifications. Record your findings.

_____

_____

12. Check the splines of the sun gear. Record your findings.

   _____

   _____

13. Sun gears should have their inner bushings inspected for looseness on their respective shafts. Record your findings.

   _____

   _____

14. Check the fit of the sun shell to the sun gear and inspect the shell for cracks, especially at the point where the gears mate with the shell. Record your findings.

   _____

   _____

15. Check the sun shell for a bell-mouthed condition where it is tabbed to the clutch drum. Any variation from a true round should be considered junk and should not be used. Record your findings.

   _____

   _____

16. Look at the tabs and check for the best fit into the clutch drum slots. This involves trial fitting the shell and drum at all the possible combinations and marking the point where they fit the tightest. Record your findings.

   _____

   _____

17. Check the gear carrier for cracks and other defects. Record your findings.

   _____

   _____

18. Check the thrust bearings for excessive wear and, if required, correct the input shaft thrust clearance by using a washer with the correct thickness. Record your findings.

   _____

   _____

19. To determine the correct thickness, measure the thickness of the existing thrust washer and compare it to the measured end play. All the pinions should have about the same end play. Record your findings.

   _____

   _____

20. Replace all defective parts and reassemble the gear set.   Task Completed ☐

Problems Encountered

_____

_____

_____

Instructor's Comments

_____

_____

_____

AUTOMATIC TRANSMISSIONS AND TRANSAXLES JOB SHEET 40

Transmission Case Service

Name _____ Station _____ Date _____

NATEF Correlation

This Job Sheet addresses the following **MAST** task:

> **C.16.** Inspect case bores, passages, bushings, vents, and mating surfaces; determine necessary action.

Objective

Upon completion of this job sheet, you will be able to inspect case bores, passages, vents, bushings, and mating surfaces.

Tools and Materials

Air nozzle   Straightedge

Crocus cloth   Feeler gauge set

Protective Clothing

Goggles or safety glasses with side shields

Describe the vehicle being worked on:

Year _____ Make _____ Model _____

VIN _____ Engine type and size _____

Transmission type and model _____

PROCEDURE

1. The transmission case should be thoroughly cleaned and all passages blown out.   Task Completed ☐

2. The passages can be checked for restrictions by applying compressed air to each one. If air flows from the other end, there is no restriction. Describe your findings.

   _____

   _____

3. To check for leaks, plug off one end of the passage and apply air to the other. If pressure builds up in that passage, there are probably no leaks in it. Describe your findings.

   _____

   _____

4. Check the fit of the servo piston in the bore without the seal to be sure it has free travel. There should be no tight spots or binding over the whole range of travel. Any deep scratches or gouges that cause binding of the piston will require case replacement. Describe your findings.

_____

_____

5. Accumulator bores are checked the same as servo bores. Describe your findings.

_____

_____

6. Check the oil pump bore at the front of the case. Describe your findings.

_____

_____

7. Case-mounted hydraulic clutch bores are prone to the same problems as servo bores. Look for any scratches or gouges in the sealing area that would affect the rubber seals. It is possible to damage these areas during disassembly, so be careful with tools used during overhaul. Describe your findings.

_____

_____

8. Sealing surfaces of the case should be inspected for surface roughness, nicks, or scratches where the seals ride. Imperfections in steel or cast-iron parts can usually be polished out with crocus cloth. Describe your findings.

_____

_____

9. Check the passages in the case for cross-tracking of one circuit to another. Fill the circuit with solvent and watch to see if the solvent disappears or leaks away. If the solvent goes down, you should check each part of the circuit to find where the leak is. Describe your findings.

_____

_____

10. Make sure all necessary check balls were in position during disassembly.   Task Completed ☐

11. Check the valve body mounting area for warpage with a straightedge and feeler gauge. This should be done in several locations. If there is a slight burr or high spot, it can be removed by flat filing the surface. Describe your findings.

   _____

   _____

12. A long straightedge should be laid across the lower flange of the case to check for distortion. Any warpage found here may result in circuit leakage, causing any number of hydraulically related problems. Describe your findings.

   _____

   _____

13. Check all bell-housing bolt holes and dowel pins. Cracks around the bolt holes indicate that the case bolts were tightened with the case out of alignment with the engine block. Describe your findings.

   _____

   _____

14. Check all of the bolts that were removed during disassembly for aluminum on the threads. If so, the thread bore is damaged and should be repaired. Thread repair entails the installation of a thread insert or by re-tapping the bore. After the threads have been repaired, make sure you thoroughly clean the case. Describe your findings.

   _____

   _____

15. The small screens found during teardown should be inspected for foreign material. Describe your findings.

   _____

   _____

16. Most screens can be removed easily. Care should be taken when cleaning because some cleaning solvents will destroy the plastic screens. Low air pressure (approximately 30 psi) can be used to blow the screens out in a reverse direction.

   Task Completed ☐

17. Bushings in a transmission case are normally found in the rear of the case and require the same inspection and replacement techniques as other bushings in the transmission. Always be sure that the oil passage to a pressure-fed bushing or bearing is open and free of dirt and foreign material. Describe your findings.

   _____

   _____

18. Vents are located in the pump body or transmission case and provide for equalization of pressures in the transmission. These vents can be checked by blowing low-pressure air through them, squirting solvent or brake cleaning spray through them, or by pushing a small diameter wire through the vent passage. Describe your findings.

_____

_____

Problems Encountered

_____

_____

_____

Instructor's Comments

_____

_____

_____

AUTOMATIC TRANSMISSIONS AND TRANSAXLES JOB SHEET 41

Servicing Internal Transaxle Drives

Name _____ Station _____ Date _____

NATEF Correlation

This Job Sheet addresses the following **MAST** task:

C.17. Inspect transaxle drive, link chains, sprockets, gears, bearings, and bushings; perform necessary action.

Objective

Upon completion of this job sheet, you will be able to inspect transaxle drive, link chains, sprockets, gears, bearings, and bushings.

Tools and Materials

Marking tool

Machinist rule

Protective Clothing

Goggles or safety glasses with side shields

Describe the vehicle being worked on:

Year _____ Make _____ Model _____

VIN _____ Engine type and size _____

PROCEDURE

1. This inspection is done with the transaxle on a bench and partially disassembled, and should be repeated as a double-check during reassembly. Begin by checking chain deflection between the centers of the two sprockets. Deflect the chain inward on one side until it is tight.   Task Completed ☐

2. Mark the housing at the point of maximum deflection.   Task Completed ☐

3. Deflect the chain outward on the same side until it is tight.   Task Completed ☐

4. Again mark the housing in line with the outside edge of the chain at the point of maximum deflection.   Task Completed ☐

5. Measure the distance between the two marks. If this distance exceeds specifications, replace the drive chain. Describe your findings.

   _____

   _____

6. Be sure to check for an identification mark on the chain during disassembly. These can be painted or dark-colored links, and may indicate either the top or the bottom of the chain, so be sure you remember which side was up. How was your chain marked?

_____

_____

7. The sprockets should be inspected for tooth wear and for wear at the point where they ride. If the chain was found to be too slack, it may have worn the sprockets in the same manner that engine timing gears wear when the timing chain stretches. A slightly polished appearance on the face of the gears is normal. Describe your findings.

_____

_____

8. Check the bearings and bushings used on the sprockets for damage. Describe your findings.

_____

_____

9. The radial needle thrust bearings must be checked for any deterioration of the needles and cage. Describe your findings.

_____

_____

10. The running surface in the sprocket must also be checked because the needles may pound into the gear's surface during abusive operation. Describe your findings.

_____

_____

11. The bushings should be checked for any signs of scoring, flaking, or wear. Describe your findings.

_____

_____

12. Based on the preceding steps, what parts need to be replaced?

_____

_____

Problems Encountered

_____

_____

_____

Instructor's Comments

_____

_____

_____

AUTOMATIC TRANSMISSIONS AND TRANSAXLES JOB SHEET 42

Servicing Final Drive Components

Name _____ Station _____ Date _____

NATEF Correlation

This Job Sheet addresses the following **MAST** task:

C.18. Inspect, measure, repair, adjust, or replace transaxle final drive components.

Objective

Upon completion of this job sheet, you will be able to inspect, measure, and adjust transaxle final drive units.

Tools and Materials

Basic hand tools

Protective Clothing

Goggles or safety glasses with side shields

Describe the vehicle being worked on:

Year _____ Make _____ Model _____

VIN _____ Engine type and size _____

Transaxle type and model _____

PROCEDURE

1. Final drive units may be helical gear or planetary gear units. A careful inspection of the assembly is done with the transaxle disassembled. The helical type should be checked for worn or chipped teeth, overloaded tapered roller bearings, and excessive differential side gear and spider gear wear. Describe your findings.

   _____

   _____

2. Measure the clearance between the side gears and the differential case. Compare your measurement to specifications. Describe your findings.

   _____

   _____

3. Check the fit of the spider gears on the spider gear shaft. Describe your findings.

_____

_____

4. Check the assembly's end play. How do your measurements compare to specifications?

_____

_____

5. What is used to preload the side bearings on this transaxle?

_____

_____

6. With a torque wrench, measure the amount of rotating torque. Compare your readings against specifications. Describe your findings.

_____

_____

7. If the bearing preload and end play is fine, as is the condition of the bearings, the parts can be reused. However, always install new seals during assembly.   Task Completed ☐

8. Planetary-type final drives are checked for the same problems as helical types. Check for worn or chipped teeth, overloaded tapered roller bearings, and excessive differential side gear and spider gear wear. Describe your findings.

_____

_____

9. The planetary pinion gears need to be checked for looseness or roughness on their shafts and for end play. Describe your findings.

_____

_____

10. Check the end play of the assembly. How did you do this and how do your measurements compare to specifications?

_____

_____

11. What is used to preload the side bearings on this transaxle?

_____

_____

12. With a torque wrench, measure the amount of rotating torque. Compare your readings against specifications. Describe your findings.

_____

_____

13. What are your conclusions about the final drive unit?

_____

_____

_____

_____

Problems Encountered

_____

_____

_____

Instructor's Comments

_____

_____

_____

AUTOMATIC TRANSMISSIONS AND TRANSAXLES JOB SHEET 43

Inspecting Apply Devices

Name _____ Station _____ Date _____

NATEF Correlation

This Job Sheet addresses the following **MAST** tasks:

C.19. Inspect clutch drum, piston, check-balls, springs, retainers, seals, and friction and pressure plates, bands and drums; determine necessary action.

C.22. Inspect roller and sprag clutch, races, rollers, sprags, springs, cages, and retainers; determine necessary action.

Objective

Upon completion of this job sheet, you will be able to inspect various apply devices of a transmission.

Tools and Materials

Compressed air and air nozzle   Lint-free shop towels

Supply of cleaning solvent   Service information

Protective Clothing

Goggles or safety glasses with side shields

Describe the vehicle being worked on:

Year _____ Make _____ Model _____

VIN _____ Engine type and size _____

Model and type of transmission _____

PROCEDURE

1. Disassemble the transmission into major units. Set each unit aside until this job sheet refers to it. Describe any problems you encountered while disassembling the transmission. Be sure to follow the procedures given in the service information while taking the transmission apart.

   _____

   _____

   _____

   _____

   _____

2. Clean each overrunning clutch assembly in fresh solvent. Allow them to air dry.

Task Completed ☐

3. Check the rollers and sprags for signs of wear or damage. Describe your findings.

_____

_____

4. Check the springs for distortion, distress, and damage. Describe your findings.

_____

_____

5. Check the inner race and the cam surfaces for scoring and other damage. Describe your findings.

_____

_____

6. Check the condition of the snap rings. Describe your findings.

_____

_____

7. Summarize the condition of the overrunning clutch units.

_____

_____

_____

_____

8. Wipe the transmission's bands clean with a dry, lint-free cloth.

Task Completed ☐

9. Check the bands for damage, wear, distortion, and lining faults. Describe your findings.

_____

_____

10. Inspect the band apply struts for damage, distortion, and other damage. Describe your findings.

_____

_____

11. Summarize the conditions of the bands.

   _____

   _____

   _____

   _____

12. Clean servo and accumulator parts in fresh solvent and allow them to air dry.   Task Completed ☐

13. Inspect their bores for scoring and other damage. Describe their condition.

   _____

   _____

14. Check the piston and piston rod for wear, nicks, burrs, and scoring. Describe your findings.

   _____

   _____

15. Inspect all springs for damage and distortion. Describe their condition.

   _____

   _____

16. Check the mating surfaces between the piston and the walls of their bores for scoring, wear, nicks, and other damage. Describe your findings.

   _____

   _____

17. Check the movement of each piston in its bore. Describe that movement.

   _____

   _____

18. Check the fluid passages for restrictions and clean out any dirt present in the passages. Describe your findings.

   _____

   _____

19. Summarize the condition of the servos and accumulators in this transmission.

   _____

   _____

Problems Encountered

_____

_____

_____

Instructor's Comments

_____

_____

_____

AUTOMATIC TRANSMISSIONS AND TRANSAXLES JOB SHEET 44

Checking and Overhauling a Multiple Friction Disc Assembly

Name _____ Station _____ Date _____

NATEF Correlation

This Job Sheet addresses the following **MAST** tasks:

C.20. Measure clutch pack clearance; determine necessary action.

C.21. Air test the operation of clutch and servo assemblies.

Objective

Upon completion of this job sheet, you will have demonstrated the ability to air test clutch packs and servos, and also disassemble, inspect, and reassemble a multiple friction disc assembly.

Tools and Materials

Feeler gauge set

Special tools for the assigned transmission

Emery cloth

Crocus cloth

Cleaning solvent

Lint-free shop towels

Air nozzle

Air test plate

Service information

Protective Clothing

Goggles or safety glasses with side shields

Describe the transmission being worked on:

Model and type of transmission _____

Vehicle the transmission is from _____

Year _____ Make _____ Model _____

VIN _____ Engine type and size _____

NOTE: *Never attempt to air-check a clutch pack until the clutch plates and snap rings are installed so as to prevent the assembly from causing personal injury or damage from overextension. Never apply more than the specified air pressure to test the transmission.*

PROCEDURE

1. With the transmission disassembled, set aside each clutch pack for inspection.

Task Completed ☐

2. Disassemble each clutch pack. Keep each assembly separate from the others. Record the number of steel and friction plates in each assembly.

_____

_____

_____

3. Check the steel plates for discoloration, scoring, distortion, and other damage. Describe their condition.

_____

_____

_____

4. If the steel plates are in good condition, rough up the shiny surface with emery (80-grit) cloth.   Task Completed ☐

5. Inspect the friction plates for wear, damage, and distortion. Describe their condition.

_____

_____

_____

6. Soak all new and reusable friction plates in ATF for at least 15 minutes before reassembling the clutch pack.   Task Completed ☐

7. Check the pressure plate for discoloration, scoring, and distortion. Describe its condition.

_____

_____

_____

8. Inspect the coil springs for distortion and damage. Describe their condition.

_____

_____

_____

9. Inspect the Belleville spring (if applicable) for wear and distortion. Make sure to check the inner fingers. Describe your findings.

_____

_____

_____

10. Check the clutch drums for damaged lugs, grooves, and splines. Also check them for score marks. Describe their condition.

   _____

   _____

   _____

11. Check the movement of the check balls in the drums. Describe your findings.

   _____

   _____

   _____

12. Summarize the condition of the clutch packs.

   _____

   _____

   _____

13. Reassemble the clutch packs with the required new parts and with new seals. Follow the recommended procedure for doing this.   Task Completed ☐

14. Insert a feeler gauge between the pressure plate and the snap ring. What is the measured clearance? _____

15. What is the specified clearance? _____

16. If the measured clearance is not the same as the specified clearance, what should you do?

   _____

   _____

   _____

17. Do what is necessary to correct the clearance.   Task Completed ☐

18. Tip the clutch assembly on its side and insert a feeler gauge between the pressure plate and the adjacent friction plate. What is the measured clearance? _____

19. What is the specified clearance? _____

20. If the measured clearance is not the same as what was specified, what should you do?

   _____

   _____

   _____

21. Do what is necessary to correct the clearance.   Task Completed ☐

22. After the clearance is set, check the service manual for the proper procedure for air testing the clutch pack. Summarize the procedure.

    _____

    _____

    _____

23. Place the pack or the partially assembled transmission in a vise or on a stand.

    Task Completed ☐

24. Apply low air pressure to the designated test port. Pay attention to the sound of the air leaking and the activation of the disc pack. Describe what you heard.

    _____

    _____

25. Release the air and listen for the release of the pack. Describe what happened.

    _____

    _____

    _____

26. Based on the preceding steps, what are your conclusions?

    _____

    _____

    _____

27. When the transmission is assembled, install the air test plate to the designated area on the transmission. What does the plate attach to?

    _____

    _____

    _____

28. What components can be checked with the test plate?

    _____

    _____

    _____

29. Apply low pressure to each of the test ports and record the results.

    _____

    _____

    _____

30. What is indicated by the results of this test?

_____

_____

_____

Problems Encountered

_____

_____

_____

Instructor's Comments

_____

_____

_____

— NOTES —

— NOTES —

— NOTES —

— **NOTES** —